Linear Electronics

RIVER PUBLISHERS SERIES IN CIRCUITS AND SYSTEMS

Series Editors:

MASSIMO ALIOTO
National University of Singapore
Singapore

KOFI MAKINWA
Delft University of Technology
The Netherlands

DENNIS SYLVESTER
University of Michigan
USA

Indexing: All books published in this series are submitted to the Web of Science Book Citation Index (BkCI), to SCOPUS, to CrossRef and to Google Scholar for evaluation and indexing.

The "River Publishers Series in Circuits and Systems" is a series of comprehensive academic and professional books which focus on theory and applications of Circuit and Systems. This includes analog and digital integrated circuits, memory technologies, system-on-chip and processor design. The series also includes books on electronic design automation and design methodology, as well as computer aided design tools.

Books published in the series include research monographs, edited volumes, handbooks and textbooks. The books provide professionals, researchers, educators, and advanced students in the field with an invaluable insight into the latest research and developments.

Topics covered in the series include, but are by no means restricted to the following:

- Analog Integrated Circuits
- Digital Integrated Circuits
- Data Converters
- Processor Architecures
- System-on-Chip
- Memory Design
- Electronic Design Automation

For a list of other books in this series, visit www.riverpublishers.com

Linear Electronics

Marcelo Sampaio de Alencar
Federal University of Bahia
Brazil

Raphael Tavares de Alencar
Institut National Polytechnique de Grenoble
France

Raissa Bezerra Rocha
Federal University of Sergipe
Brazil

Ana Isabela Cunha
Federal University of Bahia
Brazil

Published, sold and distributed by:
River Publishers
Alsbjergvej 10
9260 Gistrup
Denmark

www.riverpublishers.com

ISBN: 978-87-7022-146-7 (Hardback)
　　　978-87-7022-145-0 (Ebook)

©2020 River Publishers

All rights reserved. No part of this publication may be reproduced, stored in a retrieval system, or transmitted in any form or by any means, mechanical, photocopying, recording or otherwise, without prior written permission of the publishers.

Marcelo S. Alencar dedicates this book to his family.

Raphael T. Alencar would like to thank his amazing children, Vicente and Cora. Although these kids possess curious, sharp, minds that keep him on his toes – today he wants to thank them for keeping him happy, simply by existing.

Raissa B. Rocha dedicates the book to God, Jesus Christ and Mother Mary, and also to Wilson, Gláucia and Thiago.

Ana Isabela Cunha would like to acknowledge her beloved parents for their invaluable lessons of life.

Contents

Preface		xiii
List of Figures		xv
List of Abbreviations		xxvii
1	**Transistor Modeling for Linear Operation**	**1**
	1.1 Fundamentals on Bipolar Junction Transistors and MOS Transistors	2
	1.2 Amplification and Biasing	7
	1.3 Transistor Models for Small Signal Operation at Low Frequencies	12
	1.4 Concluding Remarks	21
2	**Linear Amplification**	**25**
	2.1 Model of a Linear Amplifier	25
	2.2 Types of Amplifiers	27
	2.3 Feedback Amplifiers	29
3	**Amplifier Circuits**	**31**
	3.1 Voltage Amplifier	31
	3.2 Current Amplifier	31
	3.3 Transconductance Amplifier	31
	3.4 Transimpedance Amplifier	31
	3.5 Gain of Amplifiers in Series	32
	3.6 Noise Figure for Series of Amplifiers	33

viii *Contents*

4 Operational Amplifiers — 35
- 4.1 Differential Amplifier — 36
  - 4.1.1 Static Condition — 37
  - 4.1.2 Dynamic Operation — 38
- 4.2 Ideal Operational Amplifier — 42
  - 4.2.1 Positive Feedback — 44
  - 4.2.2 Negative Feedback — 44
  - 4.2.3 Inverter Amplifier — 45
  - 4.2.4 Non-inverter amplifier — 46
  - 4.2.5 Adder Amplifier — 47
- 4.3 Real Operational Amplifier — 48
  - 4.3.1 Finite Gain Influence — 48
  - 4.3.2 Offset Voltage — 54
  - 4.3.3 Bias Current — 54
  - 4.3.4 Influence of temperature — 59
  - 4.3.5 Common-mode rejection ratio — 59
  - 4.3.6 Frequency Response — 61

5 Circuits with Operational Amplifiers — 67
- 5.1 Inverting Amplifier — 67
- 5.2 Non-inverting Amplifier — 68
- 5.3 Oscillators — 69
  - 5.3.1 RC Phase Shift Oscillator — 69
  - 5.3.2 Wien Bridge Oscillator — 70
- 5.4 Buffer — 71
- 5.5 Comparator — 72
- 5.6 Adder — 73
- 5.7 Subtractor — 74
- 5.8 Adder/Subtractor — 75
- 5.9 Integrator — 76
- 5.10 Differentiator — 76
- 5.11 Instrumentation Amplifier — 77
- 5.12 Shifter — 77
- 5.13 Transresistance Amplifiers — 78
- 5.14 Precision Rectifier — 79
- 5.15 Logarithmic Amplifiers — 80
- 5.16 High-impedance Differential Amplifier — 81
- 5.17 Gyrator — 82

6 Active Filters — 85
- 6.1 First-order filters — 88
  - 6.1.1 Low-pass Filter — 88
  - 6.1.2 High-pass Filter — 89
  - 6.1.3 Band-pass Filter — 90
- 6.2 Second-order Filters — 90
  - 6.2.1 Low-pass Filter — 92
  - 6.2.2 High-pass Filter — 95
  - 6.2.3 Band-pass Filter — 96
  - 6.2.4 Band-stop Filter — 99

7 Characterization of Operational Amplifiers — 101
- 7.1 Extraction of the Offset Voltage (V_{os}) — 101
- 7.2 Extraction of Total Bias Current (I_B Total) — 102
- 7.3 Extraction of Offset Current (I_{os}) — 102
- 7.4 Extraction of the Gain — 103
- 7.5 Extraction of the Common Mode Rejection Ratio (CMRR) — 103
- 7.6 Extraction of the Power Supply Rejection Ratio (PSRR) — 104
- 7.7 Extraction of the Output Swing — 104
- 7.8 Extraction of the Short-Circuit Current (I_{sc}) — 105
- 7.9 Extraction of the Supply Current — 105
- 7.10 Offset Adjustment — 105

8 Operational Amplifier Model — 107
- 8.1 Ebers-Moll Complete Model — 107
- 8.2 Using the Loop Test — 114
  - 8.2.1 The Main Parameters — 114
  - 8.2.2 The Secondary Parameters — 116
- 8.3 Basic Test Loop for Operational Amplifiers — 117
  - 8.3.1 AC Parameters — 118
  - 8.3.2 V_n Equations — 119

9 Oscillators — 121
- 9.1 Types of Oscillators — 121
- 9.2 The Ideal Oscillator — 123
- 9.3 Fundamentals of Sinusoidal Oscillators — 126
  - 9.3.1 Barkhausen Criterion — 128
- 9.4 Limiter Circuits — 131
  - 9.4.1 An Example of a Limiter Circuit Used in Oscillators — 135

x   *Contents*

- 9.5 The Wien Oscillator 140
- 9.6 LC Oscillators 146
  - 9.6.1 The Hartley Oscillator 146
  - 9.6.2 The Colpitts Oscillator 149
  - 9.6.3 The Armstrong Oscillator 150
- 9.7 The Mixer Circuit 150
  - 9.7.1 Mixer as Frequency Converter 152
  - 9.7.2 Quadratic Mixers 153
  - 9.7.3 Mixers with Proportional and Quadratic Response . 154
  - 9.7.4 Passive Mixers 155
  - 9.7.5 Active Mixers 159
- 9.8 Voltage Control Oscillator 160

10 The Phase-Locked Loop   163
- 10.1 General Description of PLL 163
  - 10.1.1 Voltage-Controlled Oscillator (VCO) ... 166
  - 10.1.2 Phase Comparator 168
  - 10.1.3 Low Pass Filter 169
  - 10.1.4 PLL Capture Range 172
  - 10.1.5 PLL Lock Range 174
- 10.2 Mathematical Model of PLL 174
  - 10.2.1 Analysis of PLL under Small Signals ... 177
- 10.3 The PLL Digital Circuit 179
- 10.4 The PLL as Frequency Synthesizer 179

11 Continuous Wave Modulation   181
- 11.1 Amplitude Modulation 184
  - 11.1.1 Amplitude Modulation – Double Side Band-Supressed Carrier (AM-DSB-SC) 185
  - 11.1.2 Amplitude Modulation – Double Side Band (AM-DSB) 190
- 11.2 AM Modulators Circuits 194
  - 11.2.1 Quadratic Modulator 195
  - 11.2.2 Modulator by Switching or Synchronous . 197
  - 11.2.3 Balanced Modulator 199
- 11.3 AM Demodulator 201
  - 11.3.1 Envelope Demodulation 202

		11.3.2 Quadratic Detector	204
		11.3.3 Synchronous Detector	205

- 11.4 Angular Modulation 206
  - 11.4.1 Narrow-Band Angle Modulator 209
  - 11.4.2 Wide-Band Angle Modulator 210
- 11.5 FM Modulator Circuits 213
  - 11.5.1 FM Wave Indirect Generation 213
  - 11.5.2 FM Wave Direct Generation 215
- 11.6 FM Demodulator Circuits 216
  - 11.6.1 FM Demodulation with PLL 216
  - 11.6.2 Frequency Discriminator 217

Appendix A: Fourier Theory   219

- A.1 Introduction 219
- A.2 The Concept of Integration 219
- A.3 Basic Fourier Analysis 220
  - A.3.1 The Trigonometric Fourier Series 221
  - A.3.2 The Compact Fourier Series 224
  - A.3.3 The Exponential Fourier Series 225
- A.4 Fourier Transform 227
  - A.4.1 Bilateral Exponential Signal 228
  - A.4.2 Transform of the Gate Function 229
  - A.4.3 Fourier Transform of the Impulse Function 230
  - A.4.4 Transform of the Constant Function 230
  - A.4.5 Fourier Transform of the Sine and Cosine Function 231
  - A.4.6 Fourier Transform of the Complex Exponential . . . 231
  - A.4.7 Fourier Transform of a General Periodic Function . 232
- A.5 Properties of the Fourier Transform 233
  - A.5.1 Linearity of the Fourier Transform 233
  - A.5.2 Scaling Property 233
  - A.5.3 Symmetry Property 234
  - A.5.4 Time Domain Shift 235
  - A.5.5 Frequency Domain Shift 235
  - A.5.6 Differentiation in the Time Domain 235
  - A.5.7 Integration in the Time Domain 236
  - A.5.8 Convolution Theorem in the Time Domain 237
  - A.5.9 Convolution Theorem in the Frequency Domain . . 238

| A.6 | Sampling Theorem | 238 |
| A.7 | Parseval's Theorem | 242 |

References   **243**

Index   **247**

About the Authors   **253**

Preface

A considerable amount of effort has been devoted, both in industry and academia, toward the design, performance analysis and evaluation of amplification schemes and filters to be used in control systems, audio, and video equipment, instrumentation and communication systems.

This book is intended to serve as a complementary textbook for courses dealing with Linear Amplification, but also as a professional book, for engineers who need to update their knowledge in the electronics, control, and communications areas.

The book is suitable for the undergraduate as well as the initial graduate levels of Electrical and Electronics Engineering courses and is useful for professionals who want to review or get acquainted with the modern exposition of the amplification theory. The book presents essential concepts in plain language and covers the most important applications of amplifier circuits.

The book has ten chapters dealing with transistor modeling, linear amplification, types of amplifiers, operational amplifiers, electronic circuits with operational amplifiers, active filters, applications and tests with operational amplifiers, oscillators, phase-locked loops, and communication circuits.

An appendix on Fourier transform and signal spectrum is included that presents the concepts of convolution, autocorrelation, and power spectral density.

List of Figures

Figure 1.1	Simplified structure and symbol for network representation of the bipolar junction transistor (BJT): (a) NPN BJT; (b) PNP BJT. The (+) sign denotes heavy doping and the (−) sign denotes light doping.	2
Figure 1.2	Simplified structure and usual symbols for network representation of the MOS field-effect transistor (MOSFET): (a) n-channel MOSFET; (b) p-channel MOSFET. The channel width and length are W and L, respectively. The (+) sign denotes heavy doping and the (−) sign denotes light doping.	4
Figure 1.3	Pinch-off phenomenon in an n-channel MOS transistor: drain end of the inversion channel in weak inversion.	5
Figure 1.4	Amplification ability of the BJT: common emitter output characteristics (i_C vs. v_{CE} (v_{EC}), parameterized by i_B), load line and output waveforms, supposing sinusoidal input. V_{CESAT} (V_{ECSAT}) is the value of v_{CE} (v_{EC}) at the onset of active region, with a typical value of 200 mV for silicon devices.	6
Figure 1.5	Amplification ability of the MOSFET: common source output characteristics (i_D vs. v_{DS} (v_{SD}), parameterized by v_{GB}), load curve and output waveforms, supposing sinusoidal input. The dashed line represents the limit between the triode region and the saturation region, corresponding to v_{DB} equal to the pinch-off voltage V_P, which depends upon gate voltage.	7
Figure 1.6	Simplified representation of a BJT common-emitter amplifier with a resistive load (R_C).	8

xvi List of Figures

Figure 1.7	Simplified representation of a MOSFET common source amplifier with an active load (diode-connected MOS transistor).	8
Figure 1.8	Quiescent value and average value (DC level) of an arbitrary signal $v_X(t)$: the DC level V_X splits the area enveloped by the signal waveform into equal areas A_1 and A_2; the quiescent value V_{XQ} is, in general, different from V_X due to system nonlinearity.	9
Figure 1.9	Linear amplification ability of a BJT for the case of small signal amplitudes. The operation is constrained to a small neighborhood of the quiescent point and output waveforms replicate input waveform.	10
Figure 1.10	Amplification failure in the BJT saturation region.	11
Figure 1.11	The importance of bias point localization for amplification, the solid line waveform represents the maximum allowable symmetric excursion of output voltage. (a) The bias point is close to the BJT saturation region, then most of the voltage signal excursion is constrained above the quiescent value; (b) the bias point is close to BJT cut-off region, then most of the voltage signal excursion is constrained below the quiescent value.	12
Figure 1.12	Translation of bias points to the origin of output current and voltage axes.	13
Figure 1.13	BJT common-emitter output characteristics (i_C vs. v_{CE}, parameterized by i_B. The zoomed-in portion illustrates the graphical meanings of the small-signal parameters.	14
Figure 1.14	BJT common emitter input characteristics (v_{BE} vs. i_B, parameterized by v_{CE}): the zoomed-in portion illustrates the graphical meanings of small-signal parameters.	15
Figure 1.15	The hybrid parameters two-port network in common emitter configuration representing the BJT for small-signal operation at low frequencies.	16

Figure 1.16	The hybrid-π parameters two-port network in common emitter configuration representing the BJT for small-signal operation at low frequencies. . . .	19
Figure 1.17	MOSFET common source output characteristics (i_D vs. v_{DS}, parameterized by v_{GS}, for constant v_{BS}): the zoomed-in portion illustrates the graphical meanings of small-signal parameters	20
Figure 1.18	MOSFET common gate characteristic for extraction of parameters K and V_{T0}: K is approximately obtained from the slope of the curve relating the square root of i_D to v_{GS} (v_{SG}) and V_{T0} is obtained from the intersection of the rectilinear portion of this curve with the voltage axis.	22
Figure 1.19	Simple procedure for the extraction of the Early voltage value V_A from the common source output characteristic.	22
Figure 1.20	Two-port network representing the MOSFET for small-signal operation at low frequency.	23
Figure 1.21	BJT common emitter amplifier (a) and its AC equivalent network (b). The BJT in (b) may be replaced by a linear two-port network (small signal model), but not in (a).	24
Figure 2.1	Model of an active two-port device (or network), which may be used to represent a series of different amplification circuits.	26
Figure 2.2	Hybrid parameter representation illustrating the operation of a linear amplifier.	27
Figure 2.3	Ideal voltage amplifier.	27
Figure 2.4	Ideal current amplifier.	28
Figure 2.5	Ideal transconductance amplifier.	28
Figure 2.6	Ideal transimpedance amplifier.	29
Figure 2.7	Negative-feedback amplifier.	29
Figure 3.1	Block diagram of a feedback voltage amplifier. . .	32
Figure 3.2	Block diagram of a feedback current amplifier. . . .	32
Figure 3.3	Block diagram of a feedback transconductance amplifier.	32
Figure 3.4	Block diagram of a feedback trans-impedance amplifier.	33

Figure 3.5	Illustration of a circuit composed of two amplifiers in series, in order to obtain a higher gain at the output.	33
Figure 4.1	Scheme for the differential amplifier circuit.	37
Figure 4.2	Small-signal equivalent circuit for a bipolar transistor in a common-emitter configuration.	38
Figure 4.3	Equivalent circuit for the bipolar transistor in common-emitter configuration, considering the approximations.	39
Figure 4.4	Rearranged bipolar transistor equivalent circuit.	39
Figure 4.5	Alternative bipolar transistor equivalent circuit.	39
Figure 4.6	Small-signal equivalent circuit for the common-emitter configuration.	40
Figure 4.7	Graphs for $(v_1 - v_2)$, v_o, and V_{c2}.	42
Figure 4.8	Symbolic representation of the operational amplifier.	43
Figure 4.9	Positive feedback amplifier circuit.	44
Figure 4.10	Negative feedback amplifier circuit.	45
Figure 4.11	Inverter amplifier circuit.	45
Figure 4.12	Non-inverter amplifier circuit.	46
Figure 4.13	Adder amplifier circuit.	47
Figure 4.14	Non-inverter circuit.	48
Figure 4.15	Inverter circuit.	50
Figure 4.16	Inverter circuit showing the OpAmp's input and output resistances.	51
Figure 4.17	Equivalent input circuit in terms of admittances.	52
Figure 4.18	Model for the OpAmp including the off-set voltage.	54
Figure 4.19	Setup to evaluate the offset voltage of the OpAmp, $R2 \gg R1$ and $R1$ have small resistance.	55
Figure 4.20	Compensating circuit to minimize the effects of the offset voltage.	55
Figure 4.21	Representation of bias currents.	56
Figure 4.22	Circuit that minimizes the error produced by the polarization currents.	57
Figure 4.23	Circuit that has the input currents represented by current sources.	57
Figure 4.24	Circuit in which the input currents are represented by voltage sources.	58

Figure 4.25	Quadripole representation of the operational amplifier.	59
Figure 4.26	Common-mode rejection ratio, for a circuit with negative feedback: (a) Input voltage applied to the inverting input; (b) Input voltage applied to the non-inverting input.	61
Figure 4.27	Bode diagrams: (a) amplitude; (b) phase.	62
Figure 4.28	Non-inverting setup. The OpAmp gain is $A(w)$.	63
Figure 4.29	Bode diagram for the closed-loop gain.	64
Figure 5.1	Inverting amplifier.	68
Figure 5.2	Non-inverting amplifier.	68
Figure 5.3	RC phase shift oscillator.	69
Figure 5.4	Wien bridge oscillator.	70
Figure 5.5	Buffer circuit.	71
Figure 5.6	Voltage comparator circuit.	72
Figure 5.7	Comparator circuit with positive output only.	73
Figure 5.8	Window comparator and it is correspondent output *versus* input graph.	73
Figure 5.9	Adder circuit.	74
Figure 5.10	Subtractor circuit, or differential amplifier.	75
Figure 5.11	Adder/subtractor circuit.	75
Figure 5.12	Integrator circuit.	76
Figure 5.13	Differentiator circuit.	77
Figure 5.14	Instrumentation amplifier signal.	78
Figure 5.15	Shifter circuit.	78
Figure 5.16	Transresistance amplifier circuit.	79
Figure 5.17	Precision rectifier circuit.	79
Figure 5.18	Alternative circuit for a half-wave precision rectifier.	80
Figure 5.19	Full-wave precision rectifier.	80
Figure 5.20	Logarithmic amplifier circuit.	81
Figure 5.21	Anti-logarithmic circuit.	82
Figure 5.22	High-impedance differential amplifier circuit.	82
Figure 5.23	Gyrator circuit.	83
Figure 6.1	Frequency response of the LPF.	86
Figure 6.2	Frequency response of the HPF.	86
Figure 6.3	Frequency response of the BPF.	86
Figure 6.4	Frequency response of the BSF.	87
Figure 6.5	General topology of 1st-order filters.	88

xx List of Figures

Figure 6.6	First-order LPF circuit.	89
Figure 6.7	First-order HPF circuit.	90
Figure 6.8	Frequency response of the BPF: associating an LPF and an HPF.	90
Figure 6.9	First-order BPF circuit.	91
Figure 6.10	Sallen-Key topology.	91
Figure 6.11	Multiple Feedback Loop circuit.	92
Figure 6.12	Second-order LPF – Sallen-Key.	93
Figure 6.13	Second-order LPF – multiple feedback circuit.	94
Figure 6.14	Second-order HPF – Sallen-Key.	95
Figure 6.15	Second-order HPF circuit – multiple feedback topology.	96
Figure 6.16	Second-order BPF circuit – Sallen-Key.	97
Figure 6.17	Second-order BPF circuit – Sallen-Key with feedback gain.	98
Figure 6.18	Second-order BPF circuit – multiple feedback topology.	98
Figure 6.19	Block diagram for a BSF.	99
Figure 6.20	Band-stop filter topology.	100
Figure 7.1	Test circuit for the extraction of DST parameters.	102
Figure 7.2	Test circuit for the extraction of DST parameters.	102
Figure 8.1	Bode plots of the voltage gain magnitude and phase.	108
Figure 8.2	Building block for simulating the slew-rate limitation.	108
Figure 8.3	Non-ideal Op-Amp macro-model.	109
Figure 8.4	μA741 Op-Amp macro-model.	113
Figure 9.1	Block diagram with overall oscillator classification.	122
Figure 9.2	(a) Symbol representing the crystal of a piezoelectric material. (b) Equivalent electrical circuit.	123
Figure 9.3	Resonant tank circuit.	124
Figure 9.4	Capacitor and inductor charge and discharge behavior in the resonant tank.	124
Figure 9.5	Energy dissipation in a feedback loopless oscillator.	126
Figure 9.6	Root arrangement of the characteristic equation: (a) ideal oscillator. (b) lossy oscillator.	126
Figure 9.7	Basic structure of a real oscillator.	127

List of Figures   xxi

Figure 9.8	Detailed structure of a real oscillator.	128
Figure 9.9	Relationship between the poles of the characteristic function and the oscillation behavior: (a) instability and gain greater than unity. (b) balance between instability and stability. (c) stability and gain less than a unit.	130
Figure 9.10	Transfer characteristic curve of an ideal limiter. . .	131
Figure 9.11	Transfer characteristic curve of an real limiter. . . .	132
Figure 9.12	Example of a limiter circuit.	132
Figure 9.13	Limiter circuit considering excitation voltage has a positive value.	133
Figure 9.14	Limiter circuit considering excitation voltage has a negative value.	133
Figure 9.15	Examples of limiting circuits and their transfer characteristic curves.	134
Figure 9.16	Example of widely used limiting circuit.	135
Figure 9.17	Linear relationship between the input voltage and the output voltage of the limiting circuit.	136
Figure 9.18	Relationship between the input voltage and the output voltage for the positive half-cycle of the input wave.	138
Figure 9.19	Transfer characteristic of the limiter circuit considering only the lower boundary.	138
Figure 9.20	Relationship between the input voltage and the output voltage for negative half-cycle of the input wave.	139
Figure 9.21	Comparison between the sine wave and its output after the limiting circuit.	140
Figure 9.22	Resulting circuit considering the negative half-cycle of the input wave.	140
Figure 9.23	General relationship between input and output voltage of the limiting circuit.	141
Figure 9.24	Electrical circuit of the Wien oscillator.	141
Figure 9.25	Electrical circuit of the Wien oscillator with limiting circuit.	144
Figure 9.26	Another setting for the electrical circuit of the Wien oscillator with limiting circuit.	145
Figure 9.27	Electrical circuit for Hartley oscillator.	146
Figure 9.28	Inductor charging rate with time constant.	148

xxii   List of Figures

Figure 9.29	Relationship between oscillator output signal and charge, discharge and phase reversal behavior.	148
Figure 9.30	Electrical circuit for Colpitts oscillator.	149
Figure 9.31	Electrical circuit for Armstrong oscillator.	151
Figure 9.32	General block diagram of a mixer circuit.	151
Figure 9.33	Block diagram of a quadratic mixer circuit.	153
Figure 9.34	Block diagram of a linear and quadratic mixer circuit.	154
Figure 9.35	Block diagram of a mixer circuit with a single diode.	155
Figure 9.36	Electrical circuit of a mixer with a diode.	156
Figure 9.37	Electrical circuit of a balanced mixer.	157
Figure 9.38	Electrical circuit of a double-balanced mixer.	158
Figure 9.39	Block diagram of an active mixer.	159
Figure 9.40	Electrical circuit of an active mixer.	160
Figure 9.41	VCO circuit.	162
Figure 10.1	Block diagram of a negative feedback system.	164
Figure 10.2	PLL basic block diagram.	164
Figure 10.3	PLL detailed block diagram with sine input and output.	165
Figure 10.4	The process of convergence of the VCO signal frequency, from f_o to f_i.	167
Figure 10.5	Two sinusoidal signals, with time or phase difference, representing the input signal in the PLL and the VCO signal.	168
Figure 10.6	Representation of the phase comparator output generation from the comparison between the sinusoid generated by the VCO and the input signal in the PLL.	168
Figure 10.7	Low pass filter circuit.	169
Figure 10.8	Relationship between the error signal and the filter output. For a higher error signal, the higher the average voltage at the filter output will be.	170
Figure 10.9	Pulse function.	170
Figure 10.10	Curve illustrating capacitor charging: (a) capacitor initially charged; (b) capacitor initially discharged.	171
Figure 10.11	Curve illustrating the capacitor discharge.	172

Figure 10.12	Relationship between the pulse, which forms the error signal, with the capacitor charging and discharging curves.	173
Figure 10.13	Curve representing various capacitor charging and discharging cycles.	173
Figure 10.14	PLL sync within capture range, that is, between the frequencies f_1 and f_2.	174
Figure 10.15	PLL sync within lock range, that is, between the frequencies F_1 and F_2.	175
Figure 10.16	Block diagram illustrating PLL in the frequency domain.	177
Figure 10.17	Block diagram illustrating the digital PLL.	179
Figure 10.18	Block diagram illustrating the PLL as frequency synthesizer.	180
Figure 10.19	Another configuration of PLL as frequency synthesizer.	180
Figure 11.1	Basic block diagram of a communications system.	182
Figure 11.2	Simultaneous transmission of signals by time-division multiplexing: (a) frequency multiplexing; (b) frequency demultiplexing.	183
Figure 11.3	Scheme for AM-DSB-SC signal generation.	185
Figure 11.4	AM-DSB-SC modulation: (a) message signal; (b) modulated signal.	186
Figure 11.5	AM-DSB-SC modulation in the frequency domain: (a) message signal spectrum; (b) modulated signal spectrum.	187
Figure 11.6	Block diagram of the synchronous demodulation.	188
Figure 11.7	Frequency spectrum illustrating baseband component filtering.	189
Figure 11.8	AM-DSB modulation: (a) message signal; (b) modulated signal.	190
Figure 11.9	AM-DSB modulated signal with modulation index equal to unity.	192
Figure 11.10	Representation of frequency domain AM-DSB modulation: (a) spectrum of the baseband signal; (b) Spectrum of the AM-DSB modulated signal.	192
Figure 11.11	Voltage x current curve characteristic of the diode or common emitter transistor.	195
Figure 11.12	Frequency spectrum of $i_c(t)$.	196

Figure 11.13	Electrical circuit of the quadratic modulator with transistor.	197
Figure 11.14	Electrical circuit of the quadratic modulator with diode.	197
Figure 11.15	Periodic pulse train.	198
Figure 11.16	Electrical circuit of the modulator by switching with diode.	199
Figure 11.17	Electrical circuit of the modulator by switching with transistor.	199
Figure 11.18	Electrical circuit of the balanced modulator with transistor.	200
Figure 11.19	Electrical circuit of the balanced modulator with transistor.	201
Figure 11.20	Block diagram of the superheterodyne receptor.	201
Figure 11.21	Electrical circuit of the tuned filter.	202
Figure 11.22	Colpitts oscillator with the mixer circuit.	203
Figure 11.23	Electrical circuit of the envelope detector.	203
Figure 11.24	Envelope detector action: (a) Signal after diode; (b) Signal with low pass filter action; (c) Demodulated signal.	204
Figure 11.25	Frequency modulated carrier.	208
Figure 11.26	(a) message signal partition in rectangular pulses; (b) representation of each rectangular pulse by an FM signal.	210
Figure 11.27	Fourier transform of the FM wave corresponding to each rectangular pulse.	211
Figure 11.28	Illustration of the sync functions generated by each rectangular pulse, from the lowest instantaneous frequency to the highest instantaneous frequency.	211
Figure 11.29	Block diagram for narrowband PM generation.	213
Figure 11.30	Block diagram for narrowband PM generation.	214
Figure 11.31	Block diagram for wide-band PM generation.	214
Figure 11.32	Block diagram of the PLL in the FM demodulation. In this case, the output of PLL is in the amplifier.	216
Figure 11.33	Block diagram of a frequency discriminator.	217
Figure 11.34	Electrical circuit of a frequency discriminator.	218
Figure A.1	Example of a periodic signal.	221
Figure A.2	Graphic of the Fourier transform of the gate function $u(t) - u(t-T)$.	229

Figure A.3	Magnitude plot for the Fourier transform of the sine function.	232
Figure A.4	A band-limited signal $f(t)$.	239
Figure A.5	Spectrum of a band-limited signal.	239
Figure A.6	Impulse train used for sampling the signal.	240
Figure A.7	The Fourier transform of an impulse train.	240
Figure A.8	Example of a sampled signal.	241
Figure A.9	Spectrum of a sampled signal.	241

List of Abbreviations

AC	Alternate Current
AM	Amplitude Modulation
AM-DSB	Amplitude Modulation Double Side Band
AM-DSB-SC	Amplitude Modulation Double Side Band – Suppressed Carrier
AR	Active Region
BJT	Bipolar Junction Transistor
BPF	Band Pass Filter
BSF	Band Stop Filter
CMRR	Common Mode Rejection Ratio
DC	Direct Current
DST	Device Subjected to Test
FDM	Frequency Division Multiplexing
FM	Frequency Modulation
HPF	High Pass Filter
IC	Integrated Circuit
IF	Intermediate Frequency
KVL	Kirchhoff's Voltage Law
LC	Indutor-Capacitor
LogAmp	Logarithmic Amplifier
LPF	Low Pass Filter
MOS	Metal-Oxide-Semiconductor
MOSFET	Metal-Oxide-Semiconductor Field-Effect Transistor
NPN	
OpAmp	Operational Amplifier
Op-Amp	Operational Amplifier
PLL	Phase Locked Loop
PM	Phase Modulation
PNP	
PSRR	Power Supply Rejection Ratio
RC	Resistor-Capacitor

RF	Radio Frequency
SNR	Signal-to-Noise Ratio
THD	Total Harmonic Distortion
VCO	Voltage Controlled Oscillator
XOR	Exclusive OR

1
Transistor Modeling for Linear Operation

Linear or ideal amplification implies that the electrical signal amplitude is scaled whereas its waveform is preserved. Rigorously speaking, the use of linear components to accomplish amplification is required. However, electronic devices, among which semiconductor devices, are far from presenting a linear relationship between their terminal currents and voltages. Linear amplification is only approximately achieved under operational constraints for network components.

Basically, in order to attain an acceptable level of linearity in amplifiers, the signal amplitudes of currents and voltages through the electronic devices should be small compared to their quiescent values. As a matter of fact, in the context of electronics, the term linear often means a first-order approximation for the relationships between electrical variables. In mathematical language, it should be rather replaced by incrementally linear.

It should be argued, why the linear approach is so broadly applied to the analysis and design of electronic amplifiers? First, assuming linear operation renders analysis and design simple tasks, since it allows employing principles and techniques that are valid only for linear networks, such as superposition, convolution, Thévenin and Norton theorems, Fourier and Laplace transforms.

Moreover, the input amplification stage of multistage amplifiers often complies with the constraints previously mentioned, thus behaving as almost linear amplifiers. Since the output signals are still small, the assumption of linear operation in the analysis or design of this class of amplifiers leads to very reliable results. Finally, even the intermediate and output amplification stages may have their operation roughly estimated by the linear assumption and further analysis such as harmonic distortion and power evaluation may complement their characterization.

2 Transistor Modeling for Linear Operation

In the following Sections, the favorite semiconductor devices employed in amplification, the bipolar junction transistor, and the MOS transistor are briefly described, biasing aspects are discussed and transistor linear models for operation with small signals at low frequencies are presented.

1.1 Fundamentals on Bipolar Junction Transistors and MOS Transistors

The solid-state electronic devices that exhibit signal amplification capabilities are the transistors (from the contraction of transfer resistors [Millman, 1972]). In the semiconductor universe, the most used type of devices are the bipolar junction transistor (BJT) and the metal-oxide-semiconductor field-effect transistor (MOSFET or MOS transistor). The former is a three-terminal component composed of three semiconductor layers: a narrow and lightly doped n-type layer between two heavily doped p-type layers (PNP transistor), or a narrow and lightly doped p-type layer between two heavily doped n-type layers (NPN transistor), as illustrated in Figure 1.1, along with the universally adopted circuit symbols. Thus, BJT consists of two opposing PN junctions.[1]

In Figure 1.1, E, B, and C are respectively the emitter, base and collector terminals in contact with the homonym layers. With junction E-B forward biased and junction B-C reverse biased, the BJT is said to operate in the active region (AR) and is able to amplify current and voltage signals. Indeed, in the

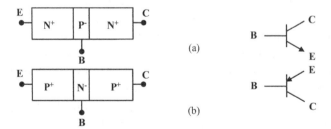

Figure 1.1 Simplified structure and symbol for network representation of the bipolar junction transistor (BJT): (a) NPN BJT; (b) PNP BJT. The (+) sign denotes heavy doping and the (−) sign denotes light doping.

---

[1] Nevertheless, two discrete diodes connected in opposition, that is, cathode with cathode or anode with anode, cannot accomplish a bipolar junction transistor, because the physical layer contact is absolutely necessary to provide the carriers and electrical field interaction that characterizes the transistor operation.

1.1 Fundamentals on Bipolar Junction Transistors and MOS Transistors

AR, the emitter (i_E) and collector (i_C) currents are amplified versions of the base current (i_B), although considerably distorted in general.

Most majority carriers of the emitter layer that diffuse towards the base layer also reach the collector layer, since they are accelerated through the reverse-biased junction by its strong electrical field. Mainly due to recombination in the narrow and lightly doped middle layer, called the base, a tiny fraction of these carriers contribute to the base current. In amplification, the variation of the base current magnitude modulates the recombination rate inside the base layer, leading to larger variations in the emitter and collector currents.

MOS transistors are four-terminal devices composed of a capacitive structure (polysilicon-oxide-silicon), bordered at each side by a PN junction, as illustrated in Figure 1.2. The polysilicon (or simply poly) layer, named gate and accessed by the gate terminal (G), represented by the gray parallelepiped in Figure 1.2, is made of a conductor material which consists of heavily doped polycrystalline silicon and performs the capacitor upper electrode [Tsividis, 1987]. Although the acronym may suggest, metal is no longer used in the construction of the device.

The silicon dioxide (SiO_2) layer, represented by the hatched parallelepiped in Figure 1.2, constitutes the capacitor insulator. The lightly doped silicon substrate acts as the lower electrode. The heavily doped layers at both capacitor sides are the source and drain diffusion layers, accessed by the source (S) and drain (D) terminals respectively. The substrate is in turn accessed by the bulk (B) terminal. Bulk potential is often constant and should assure the reverse biasing of the PN junctions in normal operation.

The MOSFET is named *n*-channel or *p*-channel whether the substrate is of the P or N-type, respectively, and the diffusion layers are accordingly of the N or P-type. Circuit symbols usually adopted for the *n*-channel and *p*-channel MOS transistors are depicted in Figure 1.2

In the enhancement MOSFET,[2] which is the most used type, provided that a drain-to-source voltage (v_{DS}) is applied, electrical current is expected to flow under the SiO_2 layer, from one of the diffusion layers to the other. However, it is only possible if an inversion channel is formed there. Such an inversion channel consists of a very thin layer of substrate minority carries (which are the majority carriers in the diffusion layers) as if the substrate type is inverted beneath the insulator.

---

[2] On the contrary of the depletion MOSFET, the enhancement MOSFET does not exhibit a pre-fabricated inversion channel.

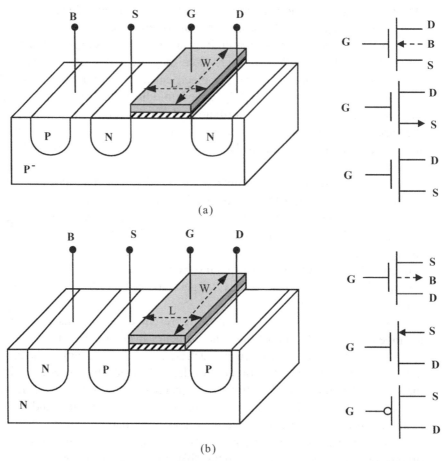

Figure 1.2 Simplified structure and usual symbols for network representation of the MOS field-effect transistor (MOSFET): (a) *n*-channel MOSFET; (b) *p*-channel MOSFET. The channel width and length are W and L, respectively. The (+) sign denotes heavy doping and the (−) sign denotes light doping.

The minority carriers are attracted to the semiconductor surface by the application of a proper gate-to-bulk voltage (v_{GB}), as the effect of the associated electrical field crossing the insulator. In the *n*-channel device, v_{GB} should be positive enough and in the *p*-channel device, it should be negative enough to obtain the required effect.

In both devices, the absolute value of v_{GB} must overcome a threshold value to provide a strong inversion level in the channel. As the absolute value

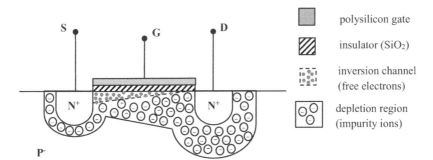

Figure 1.3 Pinch-off phenomenon in an n-channel MOS transistor: drain end of the inversion channel in weak inversion.

of v_{GB} increases, the higher is the conductivity of the inversion channel. In strong inversion, amplification of drain-to-source voltage signals happens, with respect to gate-to-source (v_{GS}) or gate-to-bulk voltage signals, only if the absolute value of the bias drain-to-bulk voltage (v_{DB}) is high enough for the channel to be weakly inverted at the drain end.

This phenomenon, illustrated in Figure 1.3, is called pinch-off and the MOSFET is said to operate in saturation since the drain current (i_D) becomes much less sensitive to drain voltage variations.

Typical output characteristics of a BJT and of a MOSFET are illustrated in Figures 1.4 and 1.5, respectively. For a BJT in common emitter mode, the output characteristics consist of the curves relating collector current i_C to collector-to-emitter voltage v_{CE}, in the case of NPN devices, or to emitter-to-collector voltage v_{EC}, in the case of PNP devices, parameterized by the base current i_B.

The common emitter mode is a conventional amplification mode for which the base is the input terminal, the collector is the output terminal and the emitter is the reference node for the alternate (AC) components of voltage signals.

The MOSFET common source output characteristics, in turn, relate drain current i_D to drain-to-source voltage v_{DS} in n-channel devices or to source-to-drain voltage v_{SD} in p-channel devices. These curves are parameterized by the gate-to-bulk voltage v_{GB} or by the gate-to-source voltage v_{GS}; the source-to-bulk voltage v_{SB} is assumed constant.

The amplification common source mode implies that the gate is the input terminal and the drain is the output terminal, with the source terminal being the reference node for the alternate (AC) components of voltage signals.

6   *Transistor Modeling for Linear Operation*

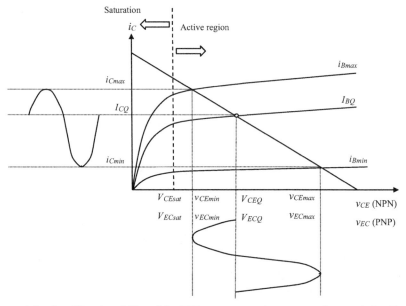

Figure 1.4 Amplification ability of the BJT: common emitter output characteristics (i_C vs. v_{CE} (v_{EC}), parameterized by i_B), load line and output waveforms, supposing sinusoidal input. V_{CESAT} (V_{ECSAT}) is the value of v_{CE} (v_{EC}) at the onset of active region, with a typical value of 200 mV for silicon devices.

The BJT and MOSFET output characteristics display an apparent similitude but it should be pointed out that:

(i) The BJT presents an exponential current-voltage relationship, whereas the current-voltage relationship in a MOSFET operating in strong inversion is quadratic (second-order polynomial)[3];
(ii) Regardless of the value of i_B, the onset of the active region in the BJT curves occurs for a constant value of v_{CE} magnitude, about 200 mV for silicon devices. The onset of saturation in the MOSFET operating in strong inversion depends on the magnitude of v_{GB}, that is, upon the inversion level.
(iii) The Early effect which consists of a slight variation in output current with output voltage in AR for the BJT and in saturation for the MOSFET has different physical origins. In the BJT, it arises from the base width modulation due to the extension of the depletion region of the

---

[3]In weak inversion, the relationship between drain current and terminal voltages is rather exponential and saturation is not related to pinch-off at the drain channel end.

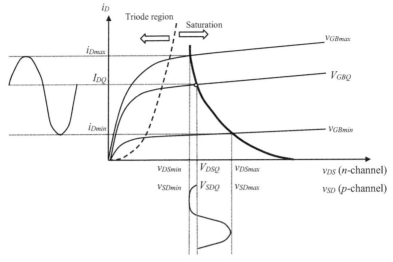

Figure 1.5 Amplification ability of the MOSFET: common source output characteristics (i_D vs. v_{DS} (v_{SD}), parameterized by v_{GB}), load curve and output waveforms, supposing sinusoidal input. The dashed line represents the limit between the triode region and the saturation region, corresponding to v_{DB} equal to the pinch-off voltage V_P, which depends upon gate voltage.

B-C junction with the increase of reverse biasing and the subsequent reduction in the recombination rate inside the base. In the MOSFET, it is a consequence of channel length modulation due to the extension of the pinched-off length at the drain channel end with the increase of reverse biasing of the D-B junction.

1.2 Amplification and Biasing

Figures 1.4 and 1.5 also show the transistor ability to amplify, since small variations in i_B (v_{GB}) in the BJT (MOSFET) could lead to large variations in i_C (i_D) or v_{CE} (v_{DS}) along the load line or load curve provided the transistor is biased in the proper region of operation, that is the active region for the BJT and saturation for the MOSFET. A load line is the graphical representation of the mesh equation obtained from the Thévenin equivalent network connected to the transistor output and reference terminals, as illustrated in Figure 1.6 in the case of BJT.

Thévenin theorem is only valid for linear networks; therefore, admit the case in which the load network is nonlinear, as depicted in Figure 1.7, in

8 Transistor Modeling for Linear Operation

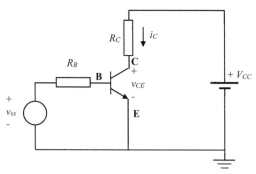

Figure 1.6 Simplified representation of a BJT common-emitter amplifier with a resistive load (R_C).

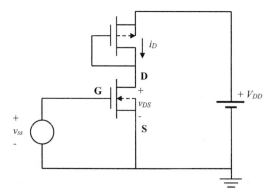

Figure 1.7 Simplified representation of a MOSFET common source amplifier with an active load (diode-connected MOS transistor).

which an active load[4] is connected to the drain terminal of the MOSFET. Thus, the relationship between the output current and the output voltage according to the nonlinear load network is graphically represented by a load curve, rather than a straight line.

In Figures 1.6 and 1.7, besides a constant voltage source, referred to as power supply, a source of time-varying voltage, referred to as voltage signal is also applied to the input terminal. The current, or voltage, at the transistor input terminal, varies continuously with time, producing a continuous change in the device output characteristic. As a result of the transistor physics, these variations are reflected in the output voltage and current, and the

---

[4] A MOSFET active load consists of a MOSFET with drain and gate terminals connected; since then the device operates in saturation and behaves as a nonlinear resistance.

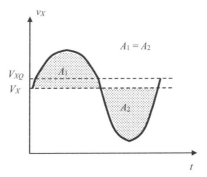

Figure 1.8 Quiescent value and average value (DC level) of an arbitrary signal $v_X(t)$: the DC level V_X splits the area enveloped by the signal waveform into equal areas A_1 and A_2; the quiescent value V_{XQ} is, in general, different from V_X due to system nonlinearity.

instantaneous operating point, (v_{CE}, i_C) for the BJT and (v_{DS}, i_D) for the MOSFET, which is the intersection between the device characteristic and the load line (curve), sweeps along the latter as the former continuously changes. The projections of this moving point on the voltage and current axes provide variations in the output voltage and current, as illustrated in Figures 1.4 and 1.5.

In the absence of an exciting signal ($v_{ss} = 0$ in Figures 1.6 and 1.7), the operating point is called the bias or quiescent point. The Q subscript is added to the variable symbols (in upper case, since they represent constant values) to denote their quiescent values, which are solely due to the power supply producing the symbols: I_{BQ}, I_{CQ}, v_{BEQ}, v_{CEQ}, v_{GBQ}, v_{DSQ}, I_{DQ}.

The current-voltage characteristics of Figures 1.4 and 1.5 are not regularly spaced parallel straight lines in the operation region. Therefore, the output projected waveforms are not scaled reproductions of the input signal, which means that they exhibit considerable distortion, often aggravated in the case of nonlinear load. For this reason, the quiescent values should not be confused with the average values (DC components), the latter separating the waveforms in half cycles of symmetric areas, as illustrated in Figure 1.8. In this case, DC components are denoted by upper case symbols without the Q subscript. Then, because of the nonlinearity: $I_C \neq I_{CQ}$, $V_{CE} \neq V_{CEQ}$, $I_D \neq I_{DQ}$ and so on.

Nevertheless, if the amplitudes of the input signals vary so little that the instantaneous operating points do not exceedingly deviate from the bias point neighborhood, the characteristics should be approximated by regularly spaced parallel straight lines, as depicted in Figure 1.9. In such cases,

10 Transistor Modeling for Linear Operation

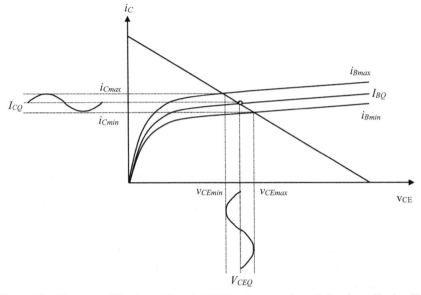

Figure 1.9 Linear amplification ability of a BJT for the case of small signal amplitudes. The operation is constrained to a small neighborhood of the quiescent point and output waveforms replicate input waveform.

distortion is not a relevant issue and output signals reproduce the input waveforms with enough reliability. The transistors are said to operate linearly from the viewpoint of the AC components of the signals.

Moreover, the average and quiescent values almost coincide so that the AC components denoted by lower case symbols with lower case subscripts are written as:

$$v_{be} = v_{BE} - V_{BE} \cong v_{BE} - V_{BEQ} \quad (1.1a)$$
$$i_b = i_B - I_B \cong i_B - I_{BQ} \quad (1.1b)$$
$$v_{ce} = v_{CE} - V_{CE} \cong v_{CE} - V_{CEQ} \quad (1.1c)$$
$$i_c = i_C - I_C \cong i_C - I_{CQ} \quad (1.1d)$$

for the BJT amplifier and

$$v_{gb} = v_{GB} - V_{GB} \cong v_{GB} - V_{GBQ} \quad (1.2a)$$
$$v_{ds} = v_{DS} - V_{DS} \cong v_{DS} - V_{DSQ} \quad (1.2b)$$
$$i_d = i_D - I_D \cong i_D - I_{DQ} \quad (1.2c)$$

for the MOSFET amplifier.

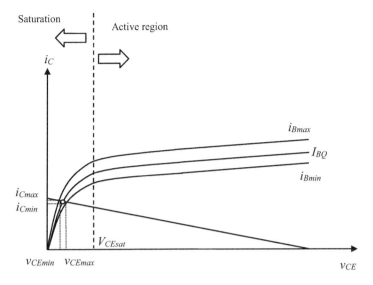

Figure 1.10 Amplification failure in the BJT saturation region.

At this point, one should be aware of the utmost importance of biasing to obtain amplification, summarized in the following observations:

(i) The power supply provides the additional energy required by the amplified signal, that is, part of the static power is converted into dynamic power to achieve the higher amplitudes at the network output.

(ii) Biasing should place the quiescent point in the proper region of operation. In the BJT (MOSFET) amplifier the large output signal variation is produced by a small variation of input signal only in the active (saturation) region, in which the characteristics are quite apart. In the saturation[5] (triode) region the characteristics are bundled up, therefore, even average variations of the parameter (input variable) lead to negligible variations of the output voltages and currents, as illustrated in Figure 1.10, which results in the absence of amplification.

(iii) Biasing should be able to place not only the quiescent point but the whole set of instantaneous operating points inside the proper region for amplification so that a large symmetric excursion of the output signal is allowed. Figure 1.11 illustrates two cases for which this condition fails.

---

[5]It is very upsetting that the region in which amplification does not occur in a BJT has the same name as the amplification region of the MOSFET.

12 Transistor Modeling for Linear Operation

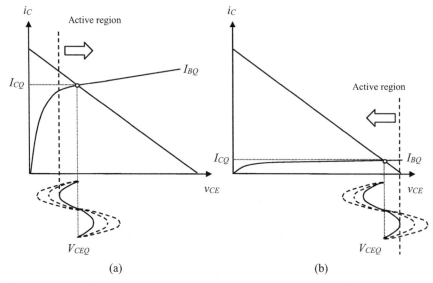

Figure 1.11 The importance of bias point localization for amplification, the solid line waveform represents the maximum allowable symmetric excursion of output voltage. (a) The bias point is close to the BJT saturation region, then most of the voltage signal excursion is constrained above the quiescent value; (b) the bias point is close to BJT cut-off region, then most of the voltage signal excursion is constrained below the quiescent value.

(iv) As discussed in the following section, the modeling parameters of transistors for small signal operation depend upon the slope and relative spacing of characteristics near the quiescent point. Since dynamic characteristics such as voltage and current gains, as well as input and output impedances depend upon these parameters, the bias point is determinant for the amplifier dynamic performance.

1.3 Transistor Models for Small Signal Operation at Low Frequencies

In this section, two BJT models and a MOSFET that is widely used for linear amplifier analysis are presented. First, it is worthwhile remarking that from (1.1) and (1.2), for small signal operation, the AC components are obtained by subtracting the quiescent values from the complete signal. Graphically, this is equivalent to shift the bias point to the origin, so that the curve corresponding to the quiescent value of the input parameter now crosses the origin, as well as the load line (curve) as illustrated in the Figure 1.12.

1.3 Transistor Models for Small Signal Operation at Low Frequencies 13

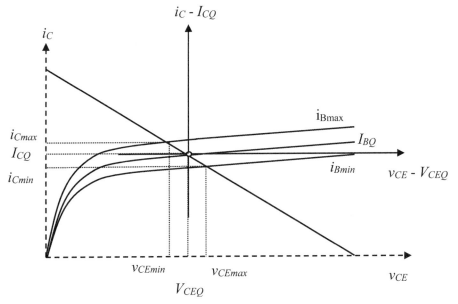

Figure 1.12 Translation of bias points to the origin of output current and voltage axes.

The axes in the curve of Figure 1.12 now represent the AC components of the output voltage (abscissa) and of the output current (ordinate). Therefore, approximate linearity is admissible only for the relationships between the AC components of the electrical signals and the linear transistor models thus derived concern uniquely the relationships between these AC components.

Indeed, the assumption of approximate linearity allows the application of the superposition theorem, even if the quiescent components of the signals do not behave linearly. Anyway, the analysis technique universally applied to solve electronic circuits operating with small signals presumes the hypothetical suppression of the power supply, therefore, only the influence of the exciting signal sources is taken into account.

This suppression is merely imaginary since without biasing there is no amplification at all[6], as pointed out in the previous section. According to this procedure, the only signals existing in the circuit are the AC components.

From now on, the constant voltage source is supposedly short-circuited and the transistor is assumed to be a linear component with the physical features provided by the biasing.

---

[6]Readers would kindly repeat this statement as a *mantra*.

14 *Transistor Modeling for Linear Operation*

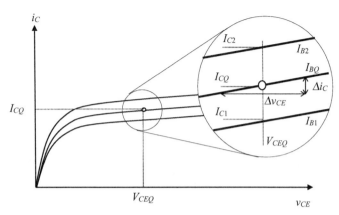

Figure 1.13 BJT common-emitter output characteristics (i_C vs. v_{CE}, parameterized by i_B. The zoomed-in portion illustrates the graphical meanings of the small-signal parameters.

a) Common emitter linear models for the BJT

According to the common-emitter characteristics of Figure 1.13, the AC component i_c of the output current can be approximately expressed as a linear function of the AC component i_b of the input current, provided the output voltage is assumed constant. Indeed, for $v_{CE} = V_{CEQ}$ and thus $v_{ce} = 0$, the ratio between any variation of i_C and the corresponding variation of i_B inside a close neighborhood of the quiescent point Q is almost constant and approaches the partial derivative of i_C with respect to i_B in the quiescent point,

$$\left.\frac{i_c}{i_b}\right|_{v_{ce}=0} \cong \left.\left(\frac{i_C - I_{CQ}}{i_B - I_{BQ}}\right)\right|_{v_{CE}=V_{CEQ}} \cong \frac{I_{C2} - I_{C1}}{I_{B2} - I_{B1}} \cong \left.\frac{\partial i_C}{\partial i_B}\right|_Q$$
(1.3a)

Moreover, the slope of the characteristic parameterized by $i_B = I_{BQ}$ is almost constant inside a close neighborhood of Q. The ratio between the AC components of i_C and v_{CE}, with $i_b = 0$, approaches this slope,

$$\left.\frac{i_c}{v_{ce}}\right|_{i_b=0} \cong \left.\left(\frac{i_C - I_{CQ}}{V_{CE} - V_{CEQ}}\right)\right|_{i_B=I_{BQ}}$$
$$\cong \left(\frac{\Delta I_C}{\Delta v_{CE}}\right)_{Qneighborhood} \cong \left.\frac{\partial i_C}{\partial v_{CE}}\right|_Q$$
(1.3b)

1.3 Transistor Models for Small Signal Operation at Low Frequencies 15

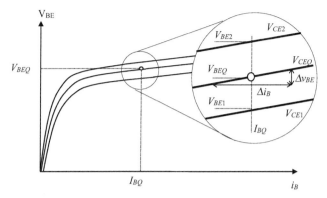

Figure 1.14 BJT common emitter input characteristics (v_{BE} vs. i_B, parameterized by v_{CE}): the zoomed-in portion illustrates the graphical meanings of small-signal parameters.

Therefore, the following approximation of i_c as a linear combination of i_b and v_{ce} applies [Millman, 1972],

$$i_c = \left.\frac{\partial i_C}{\partial i_B}\right|_Q i_b + \left.\frac{\partial i_C}{\partial v_{CE}}\right|_Q v_{ce} \quad (1.4a)$$

A similar procedure may be carried out using the common-emitter input characteristics, illustrated in Figure 1.14, leading to [Millman, 1972],

$$v_{be} = \left.\frac{\partial v_{BE}}{\partial i_B}\right|_Q i_b + \left.\frac{\partial v_{BE}}{\partial v_{CE}}\right|_Q v_{ce} \quad (1.4b)$$

The set of Equations (1.4), in which both i_c and v_{be} are expressed as linear combinations of i_b and v_{ce}, is rewritten in a more compact form as:

$$i_c = h_{fe} i_b + h_{oe} v_{ce} \quad (1.5a)$$
$$v_{be} = h_{ie} i_b + h_{re} v_{ce} \quad (1.5b)$$

in which the so-called hybrid[7] parameters in common emitter configuration are defined by the mentioned partial derivatives [Millman, 1972]:

$$h_{ie} = \left.\frac{\partial v_{BE}}{\partial i_B}\right|_Q = \left.\frac{v_{be}}{i_b}\right|_{v_{ce}=0} \quad (1.6a)$$

$$h_{re} = \left.\frac{\partial v_{BE}}{\partial v_{CE}}\right|_Q = \left.\frac{v_{be}}{v_{ce}}\right|_{i_b=0} \quad (1.6b)$$

---
[7]The term hybrid comes from the fact that they have different unities and meanings

16 Transistor Modeling for Linear Operation

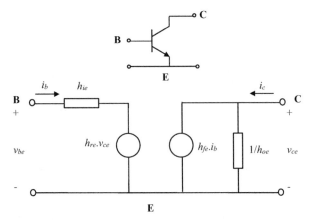

Figure 1.15 The hybrid parameters two-port network in common emitter configuration representing the BJT for small-signal operation at low frequencies.

$$h_{fe} = \left.\frac{\partial i_C}{\partial i_B}\right|_Q = \left.\frac{i_c}{i_b}\right|_{v_{ce}=0} \tag{1.6c}$$

$$h_{oe} = \left.\frac{\partial i_C}{\partial v_{CE}}\right|_Q = \left.\frac{i_c}{v_{ce}}\right|_{i_b=0} \tag{1.6d}$$

The set of Equations (1.5) generates a linear two-port network model, shown in Figure 1.15, for the bipolar junction transistor from the point of view of the small-amplitude AC components. In this model, the current summation expressed by Equation (1.5a) is represented by the parallel connection of two branches between the collector end emitter terminals, and the voltage summation expressed by Equation (1.5b) is represented by the series connection of two branches between the base and emitter terminals. The emitter terminal is common to the input and the output.

In the model of Figure 1.15, h_{ie} and h_{fe} are respectively the input impedance and the forward current gain, with short-circuited output ($v_{ce} = 0$), and h_{re} and h_{oe} are respectively the reverse voltage gain and the output admittance, with open-circuited input ($i_b = 0$). As previously described, these parameters are determined from the device characteristics at the quiescent point. Thus, the values of the model parameters change with biasing. Alternatively, they can be determined from measurements of the amplifier AC behavior, provided the circuit is biased at the desired quiescent point.

It is worthwhile noting that, for typical devices, since h_{ie} is in the order of thousands of ohms ($10^3 \Omega$) and h_{re} in the order of 10^{-4}, even considering

1.3 Transistor Models for Small Signal Operation at Low Frequencies

that v_{ce} is an amplified version of v_{be} (typically, hundreds of times the value of v_{be}), the contribution of the feedback voltage drop $h_{re}.v_{ce}$ is usually negligible in the input mesh, if compared with the voltage drop $h_{ie}.i_b$. For this reason, the parameter h_{re} is usually disregarded. On the other hand, the admittance h_{oe}, which represents the dynamic aspect of the Early effect, is often much smaller than the admittances connected between collector and emitter terminals. Whenever this applies, h_{oe} may also be disregarded. These approximations render the model still simpler.

Transistor datasheets often do not comprise the typical values of the hybrid parameters and their variation with bias current and voltage. Transistor characteristics are not frequently available either. On the other hand, circuit designers may ponder that the extraction of these parameters performing measurements could be cumbersome or not enough precise due to low-resolution equipment.

For any of these reasons, it would be useful to derive an almost equivalent AC model from the exponential characteristics of the BJT [Gummel, 1970], the hybrid-π model [Sedra, 2004]. In the active region:

$$i_C = \alpha I_{E0} e^{|v_{BE}|/(\eta \phi_t)} + I_{C0} \qquad (1.7a)$$

$$i_B = \frac{i_C - (\beta + 1) I_{C0}}{\beta} \qquad (1.7b)$$

in which α is the fraction of emitter carriers that reach the collector region, close to unit, I_{E0} and I_{C0} are the reverse saturation currents of the E-B and B-C junctions, respectively, $\phi_t = kT/q$ is the thermal voltage (k is the Boltzmann constant, T is the absolute temperature and q is the absolute value of the electron charge) and η is a fitting parameter, with a dimensionless value between 1 and 2, to take into account recombination in the depletion region of the B-C junction.

In Equation (1.7.b) the gain $\beta = \alpha/(1-\alpha)$, in the order of hundreds, is the ratio between the variations of i_C and of i_B, both referred to the cut-off condition. The value β is very close to the value of parameter $h_{FE} = i_C/i_B$ (not at all equal to h_{fe}), which is always available in datasheets. The gains $\beta \cong h_{FE}$ and α depend upon biasing point, temperature and fabrication technology.

Expression (1.7.a) relates i_C to input voltage v_{BE} and also to collector voltage through α, which is sensitive to the Early effect. Since then, the output section of this alternative model would better comprise a voltage-controlled current source, instead of the current-controlled current source of Figure 1.15.

18 Transistor Modeling for Linear Operation

Inside a close neighborhood of the quiescent point:

$$i_c = i_C - I_{CQ} \cong \left.\frac{\partial i_C}{\partial v_{BE}}\right|_Q v_{be} + \left.\frac{\partial i_C}{\partial v_{CE}}\right|_Q v_{ce} \qquad (1.8a)$$

$$i_b = i_B - I_{BQ} \cong \left.\frac{\partial i_B}{\partial v_{BE}}\right|_Q v_{be} + \left.\frac{\partial i_B}{\partial v_{CE}}\right|_Q v_{ce} \cong \left.\frac{\partial i_B}{\partial v_{BE}}\right|_Q v_{be} \qquad (1.8b)$$

In Equation (1.8.b), as previously discussed, the feedback of v_{ce} may be disregarded, leading to the right-hand side approximation. Rewriting Equations (1.8) in a more compact form, one obtains:

$$i_c = g_m v_{be} + \frac{v_{ce}}{r_o} \qquad (1.9a)$$

$$i_b = \frac{v_{be}}{r_\pi} \qquad (1.9b)$$

in which the BJT transconductance g_m, the output impedance r_o and the input impedance r_π are given by:

$$g_m = \left.\frac{\partial i_C}{\partial v_{BE}}\right|_Q \qquad (1.10a)$$

$$\frac{1}{r_o} = \left.\frac{\partial i_C}{\partial v_{CE}}\right|_Q \qquad (1.10b)$$

$$\frac{1}{r_\pi} = \left.\frac{\partial i_B}{\partial v_{BE}}\right|_Q \qquad (1.10c)$$

Comparing Equation (1.10) with Equation (1.6), a straightforward equivalence arises between r_π and h_{ie}, and between r_o and $1/h_{oe}$. Moreover, replacing $v_{be} = r_\pi \cdot i_b$ from (1.9.b) into (1.9.a) and comparing with (1.5.a) allows stating the equivalence between $g_m r_\pi$ and h_{fe}. Applying the definitions of Equation (1.10) to the large-signal model of Equation (1.7) gives [Sedra, 1998]:

$$g_m = \frac{\alpha I_{E0}}{\eta \phi_t} e^{|V_{BEQ}|/(\eta \phi_t)} \cong \frac{I_{CQ}}{\eta \phi_t} \qquad (1.11a)$$

$$r_\pi \cong \frac{\beta}{g_m} \qquad (1.11b)$$

In the set of Equations (1.11), I_{C0} has been disregarded compared with I_{CQ}. To derive Formula (1.11.b) β has been assumed independent of v_{BE} in

1.3 Transistor Models for Small Signal Operation at Low Frequencies

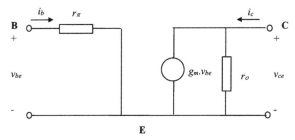

Figure 1.16 The hybrid-π parameters two-port network in common emitter configuration representing the BJT for small-signal operation at low frequencies.

the derivative of Equation (1.10.b), which leads to the incorrect equivalence between h_{fe} and β (since $h_{fe} = g_m \cdot r_\pi$).

Nevertheless, the simplicity of Equations (1.11) justify their usefulness, a rough determination of parameters g_m and r_π is possible only with the knowledge of I_{CQ} and β values (η may be assumed equal to unit for most cases). The linear two-port network representing the set of Equations (1.9) is depicted in Figure 1.16.

b) Common source linear model for the MOSFET

Differently from the BJT, for low-frequency operation, there is no input current signal in the common source configuration. This is because the gate terminal is insulated. Since the device junctions are reverse biased for normal operation, there is no current signal through the bulk terminal either. Therefore, the gate and bulk terminals are not connected to any circuit element inside the AC device models for low-frequency operation. For high-frequency operation, incremental capacitances should be added between the gate and bulk terminals and the other terminals.

The output section of the AC small-signal model may be derived from the well-known quadratic large-signal model in saturation, valid in the strong inversion region of operation [Shichman, 1968], [Tsividis, 1987]:

$$i_D = \frac{K}{2}(v_{GS} - V_T)^2 \left(1 + \frac{|v_{DS}|}{V_A}\right) \quad (1.12a)$$

$$K = \frac{\mu C'_{ox}}{n} \frac{W}{L} \quad (1.12b)$$

$$V_T = V_{T0} + (n-1)V_{SB} \quad (1.12c)$$

in which μ is the channel carrier mobility, $C'_{ox} = \epsilon_{ox}/t_{ox}$ is the oxide capacitance per unit area (ϵ_{ox} is the electric permittivity of the oxide and

20 *Transistor Modeling for Linear Operation*

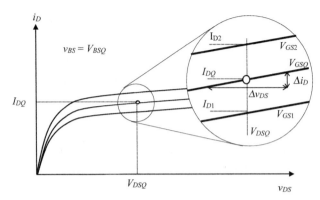

Figure 1.17 MOSFET common source output characteristics (i_D vs. v_{DS}, parameterized by v_{GS}, for constant v_{BS}): the zoomed-in portion illustrates the graphical meanings of small-signal parameters

t_{ox} is its thickness), W is the channel width and L is the channel length, V_T is the threshold voltage including the body effect (augmentation of V_T absolute value with increasing reverse biasing of the source-bulk junction), V_{T0} is the threshold voltage in equilibrium, n is the slope factor (a little above unit) and V_A is the Early voltage.

Similarly to the BJT case, according to the characteristics in Figure 1.17, inside a close neighborhood of the quiescent point,

$$\left.\frac{i_d}{v_{gs}}\right|_{v_{ds}=v_{bs}=0} \cong \left.\left(\frac{i_D - I_{DQ}}{v_{GS} - V_{GSQ}}\right)\right|_{\substack{v_{DS}=V_{DSQ}\\v_{BS}=V_{BSQ}}} \cong \left.\frac{\partial i_D}{\partial v_{GS}}\right|_Q \quad (1.13a)$$

$$\left.\frac{i_d}{v_{ds}}\right|_{v_{gs}=v_{bs}=0} \cong \left.\left(\frac{i_D - I_{DQ}}{v_{DS} - V_{DSQ}}\right)\right|_{\substack{v_{GS}=V_{GSQ}\\v_{BS}=V_{BSQ}}} \cong \left.\frac{\partial i_D}{\partial v_{DS}}\right|_Q \quad (1.13b)$$

and, although not shown in Figure 1.17,

$$\left.\frac{i_d}{v_{bs}}\right|_{v_{gs}=v_{ds}=0} \cong \left.\left(\frac{i_D - I_{DQ}}{v_{BS} - V_{BSQ}}\right)\right|_{\substack{v_{GS}=V_{GSQ}\\v_{DS}=V_{DSQ}}} \cong \left.\frac{\partial i_D}{\partial v_{BS}}\right|_Q \quad (1.13c)$$

Therefore, the AC component i_d of the drain current may be approximated by a linear combination of the AC components of the terminal voltages referred to the source terminal [Tsividis, 1987],

$$i_d = i_D - I_{DQ} \cong \left.\frac{\partial i_D}{\partial v_{GS}}\right|_Q v_{gs} + \left.\frac{\partial i_D}{\partial v_{BS}}\right|_Q v_{bs} + \left.\frac{\partial i_D}{\partial v_{DS}}\right|_Q v_{ds} \quad (1.14)$$

which is rewritten in a compact form as,

$$i_d = g_m v_{gs} + g_{mb} v_{bs} + g_d v_{ds} \tag{1.15}$$

The gate transconductance g_m, the substrate transconductance g_{mb}, and the drain conductance g_d in (1.15), which are defined by the partial derivatives in the right-hand side of (1.14), are determined through differentiation of (1.12), leading to:

$$g_m = \frac{\mu C'_{ox}}{n} \frac{W}{L} [V_{GSQ} - V_{T0} + (n-1)V_{BSQ}] \left(1 + \frac{|V_{DSQ}|}{V_A}\right)$$

$$\cong \sqrt{\frac{2\mu C'_{ox}}{n} \frac{W}{L} I_{DQ}} \tag{1.16a}$$

$$g_{mb} = (n-1)g_m \tag{1.16b}$$

$$g_d = \frac{\mu C'_{ox}}{2n} \frac{W}{L} \frac{(V_{GSQ} - V_T)^2}{V_A} \cong \frac{I_{DQ}}{V_A} \tag{1.16c}$$

In the right-hand side of Equations (1.16.a) and (1.16.c) the quiescent value of the drain current I_{DQ} has been approximated by $\frac{\mu C'_{ox}}{2n} \frac{W}{L} (V_{GSQ} - V_T)^2$, in which the Early effect has been disregarded. Parameters K, V_{T0}, and n, if not available from data sheets, may be extracted by fitting measured points to the common gate characteristic (i_D vs. v_{GS} parameterized by v_{SB}) for $v_{SB} = V_{SBQ}$ and $v_{DS} = V_{DSQ}$, as depicted in Figure 1.18.

Since g_d models the dynamic aspect of the Early effect, the Early voltage V_A is related to the slope of the output (common-emitter) characteristic corresponding to $v_{GS} = V_{GSQ}$, as illustrated in Figure 1.19. If this characteristic is virtually extended towards the left, the parameter V_A corresponds to the absolute value of its intersection with the voltage axis.

The linear network representing the MOSFET for small signal operation at low frequency is illustrated in Figure 1.20 [Tsividis, 1987]. In the very usual case in which the source and bulk terminals are connected, the current source controlled by v_{bs} is replaced by an open circuit.

1.4 Concluding Remarks

The BJT AC models introduced in the previous section represent both the PNP and the NPN types. In the same way, the MOSFET AC model applies for the *n*-channel as well as the *p*-channel devices. Since they consist of linear networks, their validity is conditioned to small-signal operation, which means

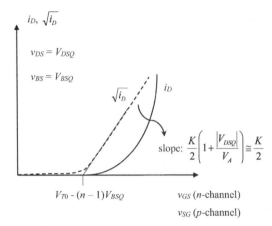

Figure 1.18 MOSFET common gate characteristic for extraction of parameters K and V_{T0}: K is approximately obtained from the slope of the curve relating the square root of i_D to v_{GS} (v_{SG}) and V_{T0} is obtained from the intersection of the rectilinear portion of this curve with the voltage axis.

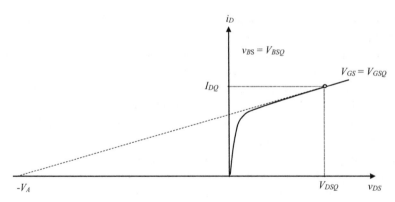

Figure 1.19 Simple procedure for the extraction of the Early voltage value V_A from the common source output characteristic.

that the AC component amplitude of each signal should be a small fraction of its DC component or its quiescent value, i.e. not more than 5 or 10%. Otherwise, the precision of the analysis may be significantly affected.

Other small-signal models, such as the T-model [Sedra, 1998] or the common-base or common-collector hybrid model for the BJT [Millman, 1972], and bulk-referred models for the MOSFET [Galup-Montoro, 2007], may be derived. Nevertheless, they are equivalent to the models here presented and may be easily converted into the others.

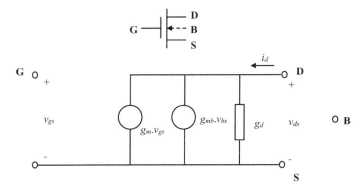

Figure 1.20 Two-port network representing the MOSFET for small-signal operation at low frequency.

It is necessary to emphasize that the transistor small-signal models represent relationships only between the AC components of the terminal voltage and current signals and thus should replace the amplifying device only in the so-called AC-equivalent network. Such a network is obtained by the virtual deactivation of the power supply in the original amplifier network.

Moreover, if there is capacitive coupling between amplifying stages or between amplifier and load or between amplifier and signal source or if there is a capacitive connection between any terminal and ground, these capacitors should be replaced by short circuits because they are designed to present negligible impedances at the operating frequencies. Figure 1.21 illustrates the AC equivalent network of a typical common-source BJT amplifier with resistive biasing.

Finally, since the models of Section 1.3 do not take into account the capacitive behavior of the semiconductor transistors, they are adequate for low-frequency operation only. Indeed, the BJT forward-biased junction E-B presents a nonlinear dynamic capacitance named diffusion capacitance, which represents the charge variation of the minority carriers at the border of the depletion region with the voltage between the E and B terminals. In the reversely biased junction B-C, in turn, the so-called transition capacitance represents the variation of the impurity ions charge with the voltage between the B and C terminals.

In the MOSFET device, dynamic intrinsic capacitances occur between all terminals, except between the drain and source terminals, because terminal voltages affect the charge distribution inside the inversion channel and inside the depletion region under the nonlinear MOS capacitor.

24   *Transistor Modeling for Linear Operation*

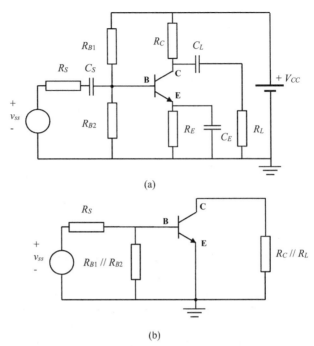

Figure 1.21   BJT common emitter amplifier (a) and its AC equivalent network (b). The BJT in (b) may be replaced by a linear two-port network (small signal model), but not in (a).

There are also extrinsic capacitances related to the reversely biased drain-bulk and source-bulk junctions and extrinsic capacitances due to the overlap between polysilicon gate and drain or source diffusion areas. Although very small, all these BJT and MOSFET incremental capacitances set an upper limit to the amplifier bandwidth, which is the range of frequencies for constant small-signal gain.

2

Linear Amplification

2.1 Model of a Linear Amplifier

A linear amplifier can be described by a two-port (or similarly, a four-terminal) device, as outlined in Figure 2.1, in which the input voltage and input current are represented, in addition to the output voltage and output current. These variables are sufficient to specify the operation of many different electronic devices, more specifically, linear devices (DeFrance, 1976).

From this representation, we can write a system of equations that mathematically models the two-port device. We need only to select two variables to be the independent variables, such as the input current (i_i) and output voltage (v_o), while the output current (i_o) and input voltage (v_i) correspond to the dependent variables. Therefore, the relation between the variables is given as (Millman and Halkias, 1972).

$$v_i = h_i \times i_i + h_r \times v_o, \tag{2.1}$$
$$i_o = h_f \times i_i + h_o \times v_o, \tag{2.2}$$

which can be represented, in matrix form, as

$$\begin{bmatrix} v_i \\ i_o \end{bmatrix} = \begin{bmatrix} h_i & h_r \\ h_f & h_o \end{bmatrix} \begin{bmatrix} i_i \\ v_o \end{bmatrix} \tag{2.3}$$

and can also be found in the literature as

$$\begin{bmatrix} v_1 \\ i_2 \end{bmatrix} = \begin{bmatrix} h_{11} & h_{12} \\ h_{21} & h_{22} \end{bmatrix} \begin{bmatrix} i_1 \\ e_2 \end{bmatrix}. \tag{2.4}$$

The four quantities h_i, h_o, h_f, and h_f are called the hybrid parameters (or *h*-parameters) of the system, since they have mixed dimensions. They represent the input impedance (in Ω), the output admittance (in S, or Ω^{-1}), the forward current gain and the reverse voltage gain (the latter two being dimensionless).

26 Linear Amplification

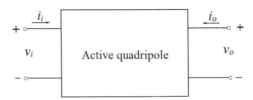

Figure 2.1 Model of an active two-port device (or network), which may be used to represent a series of different amplification circuits.

Assuming there are no reactive elements such as capacitors or inductors in the circuit, we can define the parameters as follows. Initially, applying the condition that the output port of the network is short-circuited, that is, setting the output voltage, $v_o = 0$, for the first equation of the system, one obtains

$$h_i = \frac{v_i}{i_i}, \quad \text{for } v_o = 0, \tag{2.5}$$

that is, h_i represents the input impedance of the device, when the output is short-circuited. Applying the same condition, in the second equation of the system one obtains

$$h_f = \frac{i_o}{i_i}, \quad \text{for } v_o = 0, \tag{2.6}$$

which represents the forward current gain of the two-port network, or the ratio between output current and input current when the output port is short-circuited.

Next, applying an open-circuit condition to the input of the device, that is, setting the input current to zero, for the first equation of the system,

$$h_r = \frac{v_i}{v_o}, \quad \text{for } i_i = 0, \tag{2.7}$$

thus h_r represents the ratio between the input voltage and output voltage when the input is open-circuited, namely the reverse voltage gain of the two-port network. Finally, the same condition applied to the second equation yields

$$h_o = \frac{i_o}{v_o}, \quad \text{for } i_i = 0, \tag{2.8}$$

that is, h_o represents the output admittance of the device when the input is open-circuited, given in siemens (S) or mho (Ω^{-1}).

The relations between the currents, voltages and hybrid parameters can be seen in Figure 2.2, which shows the equivalent circuit representation of

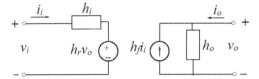

Figure 2.2 Hybrid parameter representation illustrating the operation of a linear amplifier.

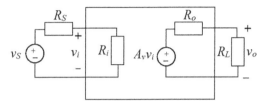

Figure 2.3 Ideal voltage amplifier.

a given two-port device (such as an amplifier), using Thévenin and Norton equivalent circuits, with controlled voltage and current sources.

For example, it can be seen that the input impedance is obtained as the ratio between the input voltage and input current, for the case when the controlled voltage source on the left part of the circuit is replaced by a wire, corresponding to the output voltage set to zero ($v_o = 0$). Other hybrid parameters can be obtained in a similar manner.

The hybrid parameter representation is useful in the analysis of linear electronic circuits, such as those with transistors or operational amplifiers.

2.2 Types of Amplifiers

In a general manner, amplifiers can be classified as signal amplifiers or power amplifiers. Power amplifiers aim to couple the maximum energy to the next amplification stage, or to a transmitter, for example. Therefore, its output impedance must be matched to the input of the subsequent stage so that the maximum power can be transferred to the load (Padilha, 1993).

On the other hand, signal amplifiers aim to couple the voltage or current signal to the next stage, for this, they operate with unmatched impedances for the signal to be completely transferred. In general, signal amplifiers can be classified in four categories:

1. Voltage Amplifiers, in which the input voltage is amplified to provide an output voltage.

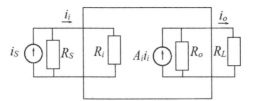

Figure 2.4 Ideal current amplifier.

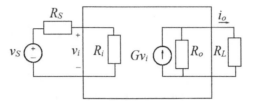

Figure 2.5 Ideal transconductance amplifier.

The ideal voltage amplifier has an infinite input impedance ($R_i \to \infty$) so the same voltage from the signal source appears at the input, that is, so that $v_i = v_S$. The output impedance, R_o, is zero, so that the voltage is completely transmitted to the load, i.e., $v_o = A_V \cdot v_i$.

2. Current Amplifiers, in which the output current is an amplified version of the input current.

   The ideal current amplifier has zero input impedance ($R_i = 0\,\Omega$) so that the same current from the signal source appears at the input, that is $i_i = i_S$. And the output impedance, R_o, is infinite so the current can be completely transmitted to the load, i.e., $i_o = A_I \cdot i_i$.

3. Transconductance Amplifiers, in which a voltage input produces a current output.

   The ideal transconductance amplifier has infinite input impedance ($R_i \to \infty$) so that the same voltage from the signal source appears at the input, that is, $v_i = v_S$. The output impedance, R_o, is also infinite so the current can be completely transmitted to the load, i.e., $i_o = G \cdot v_i$.

4. Transimpedance Amplifiers, in which the input current produces an output voltage.

   The ideal transimpedance has zero input impedance ($R_i = 0\,\Omega$) so that the same current from the signal source appears at the input, that is, $i_i = i_S$. The output impedance is also zero, so that the voltage is completely transmitted to the load, i.e., $v_o = R \cdot i_i$.

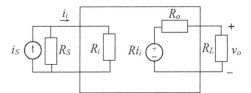

Figure 2.6 Ideal transimpedance amplifier.

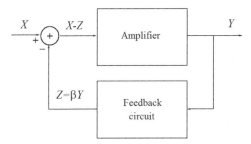

Figure 2.7 Negative-feedback amplifier.

2.3 Feedback Amplifiers

Negative-feedback circuits are used to stabilize the operation of amplifiers. All types of amplifiers can be improved with adequate use of feedback (Alley and Atwood, 1973).

For example, the input impedance of a voltage amplifier, which is usually high, can be increased through the use of feedback, while the output impedance, which should be low, can be reduced. In addition, the amplifier gain can be stabilized against variations in the hybrid parameters of transistors, or of other active devices used in the amplifier.

Moreover, negative feedback improves the amplifier's frequency response and linearity. However, these results are obtained at the expense of reducing the amplifier gain. In certain circumstances, feedback can cause instability and oscillation.

A basic scheme of an amplifier with negative feedback can be seen in Figure 2.7. The input signal, X, produces an output signal $Y = A(X - Z)$, in which A represents the amplifier's open-loop gain. The open-loop gain of an ideal operational amplifier is infinite, in practice, it is very high, in the order of 10^6.

The signal is processed through the feedback loop, producing a signal $Z = \beta Y$, in which β corresponds to the closed-loop gain, which is external

30 Linear Amplification

to the amplifier and can be controlled as (this is discussed in more detail in Chapter 3). Therefore, the differential input received by the amplifier is $X - Z = X - \beta Y$. Consequently, the final gain, A_F, for the negative-feedback amplifier is given by

$$A_F = \frac{Y}{X} = \frac{A}{1 + \beta A}. \tag{2.9}$$

One may notice that the feedback gain is lower than the original amplifier gain, by a factor of $1 + \beta A$. This also reduces the sensitivity of the amplifier by the same factor. Furthermore, it is usually true that the product $\beta A \gg 1$, which makes the circuit's final gain virtually independent from the amplifier's open-loop gain

$$A_F = \frac{1}{\beta}. \tag{2.10}$$

3
Amplifier Circuits

3.1 Voltage Amplifier

The voltage amplifier operates by feeding back the voltage output, generally collected at the load terminals. The signal from the feedback loop, which is a voltage proportional to the output, is inserted in series with the input signal. This causes the input impedance to increase proportional to the closed-loop gain, β, and the output impedance to reduce in the same proportion (Gronner, 1976).

3.2 Current Amplifier

The current amplifier operates by feeding back the output current, which is generally the same current that goes through the load (Mayaram, 2008). The signal from the feedback loop, a current proportional to that flowing at the output, is inserted in parallel with the input signal. This causes the input impedance to be reduced proportional to the closed-loop gain, β, and the output impedance to increase in the same proportion.

3.3 Transconductance Amplifier

The transconductance amplifier feeds back the output current. The signal from the feedback loop, a voltage proportional to the output current, is inserted in series with the input signal. This causes the input and output impedances to increase proportionally to the closed-loop gain (Millman and Halkias, 1972).

3.4 Transimpedance Amplifier

The transimpedance amplifier feeds back the output voltage collected at the load terminals. The signal from the feedback loop, a current proportional to

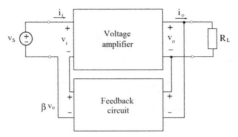

Figure 3.1 Block diagram of a feedback voltage amplifier.

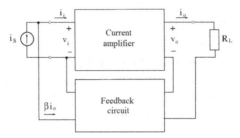

Figure 3.2 Block diagram of a feedback current amplifier.

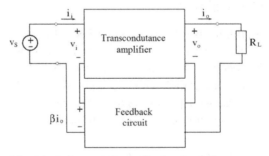

Figure 3.3 Block diagram of a feedback transconductance amplifier.

the output voltage, is inserted in parallel to the input current. This causes the input and output impedances to be reduced proportionally to the closed-loop gain (Padilha, 1993).

3.5 Gain of Amplifiers in Series

Coupling amplifiers in series allows the circuit to increase its gain, either allowing impedance matching in the case of power amplifiers; or impedance mismatching in the case of signal amplifiers (Silva, 2013).

3.6 Noise Figure for Series of Amplifiers

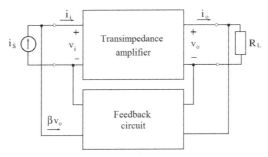

Figure 3.4 Block diagram of a feedback transimpedance amplifier.

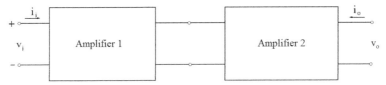

Figure 3.5 Illustration of a circuit composed of two amplifiers in series, in order to obtain a higher gain at the output.

For the circuit shown in Figure 3.5, if we consider the amplifiers have voltage gains G_1 and G_2, for example, the total gain is $G = G_1 \times G_2$. The same occurs for current amplifiers.

3.6 Noise Figure for Series of Amplifiers

Coupling also allows the circuit to compensate for the internal noise from the amplifiers, represented by its figure noise, F. When two amplifiers, with noise figures F_1 and F_2 and gains G_1 and G_2, are series-coupled (or cascaded), the total noise figure is given by

$$F = F_1 + \frac{F_2}{G_1}. \qquad (3.1)$$

Therefore, it suffices that the first amplifier has a low noise figure, in addition to a high gain, for the total noise figure to be practically the same as the first stage.

This strategy is largely applied for the reception of satellite signals, which are very weak as they reach the Earth's surface. A Low Noise Amplifier (LNA) is commonly found installed directly on the reception antenna.

4
Operational Amplifiers

The operational amplifier, colloquially known as OpAmp (or op-amp), was originally developed to be used in analog computers. The name of the circuit itself stems from the fact that it could be connected so to procure a large variety of mathematical operations, among these are inversion, addition, subtraction, differentiation, and integration (Alley and Atwood, 1973).

However, at a first instance, any OpAmp is simply a high-gain amplifier with DC coupling and can, therefore, be used in any application in which one may desire to amplify a voltage signal. As a circuit with lumped elements, the OpAmp was never popular, its cost was impractically high, besides the fact that it showed stability issues due to the imperfect matching between the transistors in the input stage. The circuit integration onto a single substrate solved both of these problems. Amplifiers in monolithic ICs are low-cost, possess quasi-perfect matching of the transistors and procure good thermal stability.

In terms of functionality, the OpAmp circuit can be divided into three parts:

1. The input section is composed of a *High-Impedance Differential Amplifier*. With bipolar transistors in the initial stage, the input impedance can vary from 10 kΩ to 100 kΩ. For higher impedances (up to 1 MΩ), the Darlington configuration can be used on each side of the differential amplifier (Cutler, 1977). With the use of Field-Effect Transistors (FETs), input impedances in the order of 100 MΩ can be obtained.
2. The second section is designed to provide the full gain of the OpAmp. Frequently, it also consists of a differential amplifier. The external terminals for controlling the frequency and phase responses are normally connected to this section and are used to establish the conditions for each specific application.

36 Operational Amplifiers

3. The final section generally consists of a power stage with low output impedance. There might be a simple, or push-pull, emitter follower (class A) or a class B power stage (DeFrance, 1976). These circuits tend to maintain the gain and the frequency response independent from the load. Additionally, this stage introduces voltage-level translation and can provide protection against short-circuiting at the output. The level translation is necessary for a DC amplifier, to remove the continuous component at the output, shifting the output voltage in relation to the circuit's common ground.

4.1 Differential Amplifier

Understanding the ideal OpAmp model, as well as, the real OpAmp necessarily involves the study of the differential configuration.

The differential amplifier consists of an active circuit that amplifies the difference between two input signals (Kaufman and Wilson, 1984). A typical differential amplifier setup can be seen in Figure 4.1 (the output stage used in this example is an emitter follower), the loading effect – voltage variation due to the presence of the load – is small due to the high input impedance of the circuit and small output impedance.

The circuit around transistor $T3$ works as a current source, in which $I_c = 2I_0$, and I_0 represents the quiescent current (baseline, or leakage, current present when the circuit is not amplifying a signal or driving a load) from $T1$ and $T2$.

> **Question: Prove the following results.**
>
> (a) The circuit presents a high impedance relative to the emitters of transistors $T1$ and $T2$.
> (b) Transistor $T3$ acts as a constant and temperature-independent current source if the base-emitter voltage, $V_{be3} = (V_d \cdot R5)/(R4 + R5)$.
> (c) In this case, $I_c = 2I_0 = \frac{R4 \cdot V}{R4+R5}$.
> (d) Finally, it can be proven that the common-mode gain (defined further in this chapter) is inversely proportional to the effective impedance at the emitter.

In the next section, an analysis of the differential amplifier is performed, in static condition and in dynamic operation.

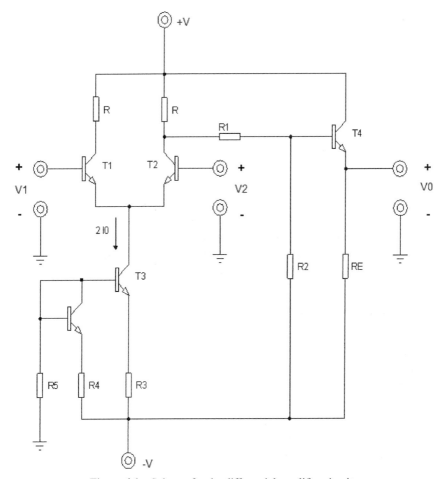

Figure 4.1 Scheme for the differential amplifier circuit.

4.1.1 Static Condition

Admitting that transistors $T1$ and $T2$ are identical, the collector currents will be equal, $I_{c1} = I_{c2} = I_0$, when $V1 = V2$.

If I_0 is set so that $V_{c2} = \frac{V}{2}$ (static condition), it is easy to determine $R1$ and $R2$ such that $V0 = 0$ V (static condition). For this, it suffices to show that

$$V0 = E_{c2} - R_1.I_{R1} = 0 \qquad (4.1)$$

The last equation is obtained thanks to the unit voltage gain from the common-collector configuration.

But
$$E_{c2} = \frac{V}{2} \tag{4.2}$$

and
$$I_{R1} = \frac{V_{c2} - V}{R_1 + R_2} = \frac{3 \times V}{2(R_1 + R_2)} \tag{4.3}$$

Thus
$$R_1 = \frac{(R_1 + R_2)}{3} \tag{4.4}$$

and
$$R_2 = 2.R_1 \tag{4.5}$$

Usually, $R1$ and $R2$ are selected so that $R1 + R2 > R$, (e.g.: $R1 + R2 = 10 \cdot R$), in which R is a known resistance value. It can be easily seen from Equation (4.1) that for V_{c2} varying between V and zero, the output voltage $V0$ will vary between $+V/3$ and $-V/3$.

4.1.2 Dynamic Operation

The small-signal equivalent circuit for the bipolar transistor in the common-emitter configuration is presented in Figure 4.2, making use of the h-parameter representation.

Considering the Early effect is negligible ($h_{re} \approx 0$), and considering a very high output impedance ($h_{oe} \approx 0$), the simplified equivalent circuit of Figure 4.3 is obtained. And by modifying the disposition of the elements, the circuit from Figure 4.4 is obtained.

Transposing the resistance h_{ie} to the emitter circuit, the scheme from Figure 4.5 is obtained, which will be used as the equivalent circuit model for transistors $T1$ and $T2$. This last passage is left for the reader.

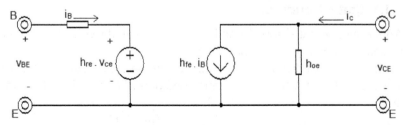

Figure 4.2 Small-signal equivalent circuit for a bipolar transistor in a common-emitter configuration.

4.1 *Differential Amplifier* 39

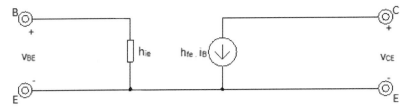

Figure 4.3 Equivalent circuit for the bipolar transistor in common-emitter configuration, considering the approximations.

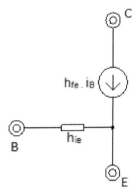

Figure 4.4 Rearranged bipolar transistor equivalent circuit.

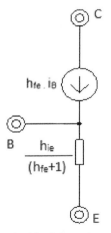

Figure 4.5 Alternative bipolar transistor equivalent circuit.

40 Operational Amplifiers

Obs.: The parameters from the model shown in Figure 4.2 are defined (Millman and Halkias, 1972):

- Input impedance in common-emitter configuration (typical value: $h_{ie} = 1.1 \text{ k}\Omega$)

$$h_{ie} = \left. \frac{V_{be}}{I_b} \right|_{Vce=0} \quad (4.6)$$

- Forward current gain (typical value: $h_{fe} = 50$)

$$h_{fe} = \left. \frac{I_c}{I_b} \right|_{Vce=0} \quad (4.7)$$

- Output admittance (typical value: $h_{oe} = 24 \text{ } \mu\text{S}$)

$$h_{oe} = \left. \frac{I_c}{V_{ce}} \right|_{Ib=0} \quad (4.8)$$

- Reverse voltage gain (typical value: $h_{re} = 2,5 \times 10^{-4}$)

$$h_{re} = \left. \frac{V_{be}}{V_{ce}} \right|_{Ib=0} \quad (4.9)$$

To determine the output voltage in dynamic operation, the small-signal equivalent circuit seen in Figure 4.6 is used. It is important to notice that

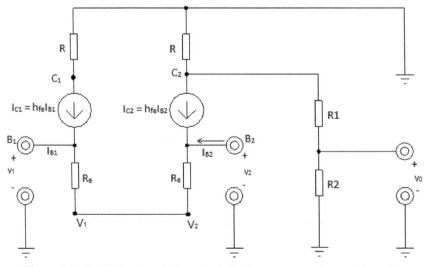

Figure 4.6 Small-signal equivalent circuit for the common-emitter configuration.

transistor $T4$ in Figure 4.1 is an emitter follower (voltage gain equal to 1), and that v_o can be calculated at either the terminals of resistors RE or $R2$.

$$i_{b1} = \frac{i_{e1}}{\beta + 1} = \frac{v_1 - v_2}{(\beta + 1) \times 2R_e} \quad (4.10)$$

$$i_{b2} = \frac{i_{e2}}{\beta + 1} = \frac{v_2 - v_1}{(\beta + 1) \times 2R_e} = -i_{b1} \quad (4.11)$$

$$V_{c2} = -\beta i_{b2} \cdot R(R_1 + R_2) \quad (4.12)$$

$$V_{c2} = -\beta \times R \times \frac{v_2 - v_1}{(\beta + 1) \times 2R_e} \quad (4.13)$$

$$V_{c2} = -\frac{R}{2R_e} \times (v_2 - v_1) = A_d(v_2 - v_1) \therefore A_d = -\frac{R}{2R_e} \quad (4.14)$$

in which A_d represents the differential stage gain, which is usually high.
The output voltage, v_o, suffers an attenuation of

$$\frac{R_2}{R_1 + R_2} = \frac{2}{3} \quad (4.15)$$

and is equal to

$$v_o = -\frac{2}{3} \times \frac{R}{2R_e}(v_2 - v_1) \quad (4.16)$$

$$v_o = -A \times (v_2 - v_1), \text{ in which the gain is} A = \frac{2}{3}A_d \quad (4.17)$$

The following is an analysis of v_o for the various possibilities of the input signals v_1 and v_2:
- $v_1 \neq v_2 \neq 0$, the output signal e_o will be equal to a positive constant multiplied by the difference between the two input signals, and if $v_1 = v_2$, then the output voltage is always zero.
- $v_1 = 0$ and $v_2 \neq 0$, the output signal v_o suffers a phase inversion, of 180°, relative to the input signal v_2.
- $v_1 \neq 0$ and $v_2 = 0$, the output signal v_o will have the same phase as the input signal v_1.

The graphs for v_o, $(v_1 - v_2)$ and V_{c2} as a function of time can be seen in Figure 4.7.
It is convenient to notice that the role of impedances $R1$ and $R2$ is to produce a potential translation of v_o, whose instantaneous value will be zero

42 Operational Amplifiers

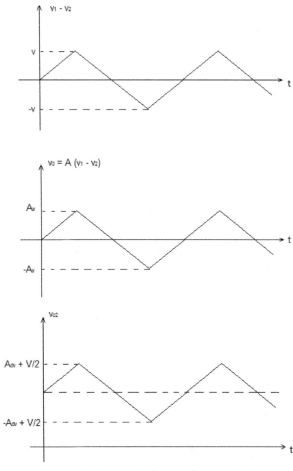

Figure 4.7 Graphs for $(v_1 - v_2)$, v_o, and V_{c2}.

whenever $v_1 = v_2$, nonetheless, their presence attenuates the global gain. One way to avoid this attenuation and still achieve the desired translation is to utilize a Zener diode of $\frac{V}{2}$ volts instead of resistor $R1$. The Zener diode will work as a voltage source of negligible internal resistance.

4.2 Ideal Operational Amplifier

The calculations for the circuit configurations employing OpAmps as active elements are relatively simple, as long as the simplified model for the amplifier in question is used. It is convenient to start the study considering

4.2 Ideal Operational Amplifier

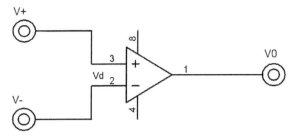

Figure 4.8 Symbolic representation of the operational amplifier.

the OpAmp as ideal, indicating its characteristics and analyzing some of its applications, and gradually, as we progress in its comprehension, certain elements will be introduced in order to explain its real operation (Padilha, 1993).

The ideal operational amplifier can be considered as a differential DC amplifier with the following characteristics:

- Infinite voltage gain ($A = \infty$);
- Infinite input resistance ($R_i = \infty$);
- Zero output resistance ($R_o = 0$);
- It is perfectly balanced ($v_o = 0$ whenever $v^+ = v^-$);
- Infinite bandwidth.

The symbolic representation of the OpAmp can be seen in Figure 4.8, in which the signs (+) and (−) correspond to the non-inverting input and inverting input, respectively, and indicate the following: if input (−) is grounded and a signal is applied to input (+), the output will be in phase with the input signal; and if a signal is, in turn, applied to input (−) and input (+) is ground, the output signal will be shifted 180° relative to the input signal.

if $v^+ = v^-$, then $v_o = 0$
if $v^- = 0$, then $v_o = A \cdot v^+$ (v_o is in phase with v^+)
if $v^+ = 0$ then $v_o = -A \cdot v^-$ (v_o is shifted 180° relative to v^-)

Considering the OpAmp as ideal, its voltage gain is infinite, therefore, whatever is the desired output voltage, it is necessary a zero voltage V_d to produce it, since

$$V_d = \frac{v_o}{A} \to 0, \quad \text{when } A \to \infty \tag{4.18}$$

One can, therefore, consider, for the amplifier gain calculation applying negative feedback, that $v^+ = v^-$. However, it should not be forgotten that this hypothesis is only valid when negative feedback is employed.

44 Operational Amplifiers

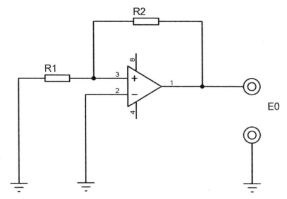

Figure 4.9 Positive feedback amplifier circuit.

4.2.1 Positive Feedback

An OpAmp with positive feedback can be seen in Figure 4.9. It can be seen that a fraction of the voltage from the output is being looped back into the non-inverting input through the use of a resistive voltage divider composed of $R1$ and $R2$. To understand the operation of the circuit in these conditions, consider that a given disturbance causes the output voltage v_o to increase, this will cause an increase in v^+ which in turn will reinforce the increase in v_o and so on, until the amplifier becomes saturated, in these conditions one obtains:

$$v^- = 0 \tag{4.19}$$

$$v^+ = \left(\frac{R1}{R1 + R2} \right) v_o \tag{4.20}$$

$$v_o = +V \tag{4.21}$$

in which $+V$ corresponds to the voltage value inputted into the amplifiers positive supply rail (node 8 in Figure 4.9).

Similarly, if v_s is negative, the output will saturate at its negative supply voltage value.

4.2.2 Negative Feedback

If negative feedback is employed, as seen in Figure 4.10, the circuit will be stable, any disturbance forcing the output v_o to increase will also produce an increase in the inverting input, which tends to eliminate the increase at the output, quickly stabilizing the output of the circuit.

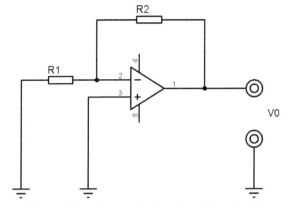

Figure 4.10 Negative feedback amplifier circuit.

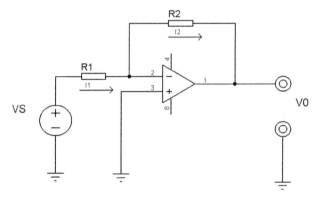

Figure 4.11 Inverter amplifier circuit.

In this case,

$$v^- = \left(\frac{R1}{R1+R2}\right)v_o \qquad (4.22)$$

4.2.3 Inverter Amplifier

Consider the negative feedback circuit seen in Figure 4.11, in order to determine the relationship between the output voltage, v_o, and the input voltage, v_s. The amplifier is considered to be ideal and we can adopt the following sequence of calculations:

(1) $\quad v^+ = v^- = 0 \qquad (4.23)$

46 Operational Amplifiers

(2) $I1 = \dfrac{v_s}{R1}$ (4.24)

(3) $I2 = I1, \quad (R_i = \infty)$ (4.25)

(4) $v_o = -R2 \cdot I2$ (4.26a)

$\ v_o = -R2 \times \dfrac{v_s}{R1}$ (4.26b)

whence

$$\dfrac{v_o}{v_s} = -\dfrac{R2}{R1} \qquad (4.27)$$

The circuit's gain is negative and its value is determined uniquely from the ratio between resistances $R2$ and $R1$. The inconvenient aspect of this circuit is that the impedance seen from the signal source is the resistance $R1$ itself, whose value should not be too high in order to minimize the effect of the bias current.

4.2.4 Non-inverter amplifier

The circuit from Figure 4.12 can be analyzed following the same steps previously described. It can be seen that, besides not inverting the signal's phase, the impedance seen from the signal source is very high, since the source terminal is connected directly to the amplifier input. This circuit is recommended when aiming to cause the least impact on the signal source (Silva, 2013).

(1) $v^+ = v^- = v_s$ (4.28)

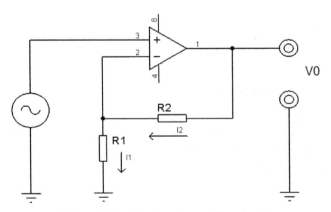

Figure 4.12 Non-inverter amplifier circuit.

(2) $I_1 = \dfrac{v_s}{R_1}$ (4.29)

(3) $I_1 = I_2$ (4.30)

(4) $v_o = (R_1 + R_2) \cdot I_2$ (4.31)

$$v_o = (R_1 + R_2) \times \dfrac{v_s}{R_1} \quad (4.32)$$

$$v_o = \left(1 + \dfrac{R_2}{R_1}\right) v_s \quad (4.33)$$

$$\dfrac{v_o}{v_s} = 1 + \dfrac{R_2}{R_1} \quad (4.34)$$

4.2.5 Adder Amplifier

To obtain a signal which is proportional to the sum of several other signals, the adder amplifier circuit seen in Figure 4.13 can be used. Following the same steps as with the previous circuits:

(1) $v^+ = v^- = 0$ (4.35)

(2) $I_1 = \dfrac{v_1}{R_1}, \; I_2 = \dfrac{v_2}{R_1}, \; I_3 = \dfrac{v_3}{R_1}$ (4.36)

(3) $I_T = I_1 + I_2 + I_3$ (4.37)

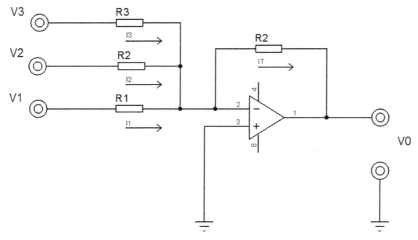

Figure 4.13 Adder amplifier circuit.

(4) $\quad v_o = -R_2 I_T$ (4.38a)

$$v_o = -R_2 \left(\frac{v_1}{R_1} + \frac{v_2}{R_2} + \frac{v}{R_3} \right), \quad \text{for } R_1 = R_2 = R_3 \qquad (4.38b)$$

$$v_o = -\frac{R_2}{R_1}(v_1 + v_2 + v_3) \qquad (4.38c)$$

A variety of circuits that use OpAmps is shown in Chapter 5.

4.3 Real Operational Amplifier

Henceforth, the influence of elements that were previously put aside are analyzed, considering an ideal amplifier model. That must be taken into account in order to explain the OpAmp practical operation. These elements are: bias current, offset current, offset input voltage, finite gain, gain as a function of frequency, finite input resistance, non-zero output resistance and common-mode rejection ratio (CMRR).

4.3.1 Finite Gain Influence

In order to evaluate the effect of the OpAmp's finite open-loop gain (A) over the closed-loop gain (G), a simple analytical expression relating A and G must be established.

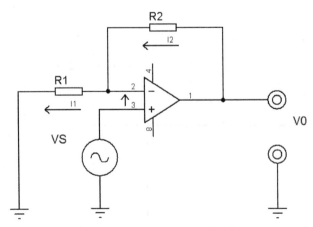

Figure 4.14 Non-inverter circuit.

Non-inverter circuit
The following equations can be obtained from the circuit:

$$I_1 = \frac{v_s + v_d}{R_1} \tag{4.39}$$

$$I_2 = \frac{v_o - (v_s + v_d)}{R_1} \tag{4.40}$$

$$v_o = -v_d \times A. \tag{4.41}$$

If the OpAmp's input resistance is considered infinite, $I_1 = I_2$ and, as a function of v_o and v_s, it can be seen that

$$v_d = \frac{R_1}{R_1 + R_2} \times v_o - v_s, \tag{4.42}$$

Therefore,

$$v_o = -A \cdot \left(\frac{R_1}{R_1 + R_2} \times v_o - v_s \right), \tag{4.43}$$

and defining

$$\beta = \frac{R_1}{(R_1 + R_2)} \tag{4.44}$$

yields

$$\frac{v_o}{v_s} = \frac{A}{1 + \beta A} = \frac{1}{\beta} \times \frac{1}{1 + 1/(\beta A)}. \tag{4.45}$$

Considering the geometric power series expansion for the fraction

$$\frac{1}{1+x} = 1 - x + \frac{x^2}{2} - \frac{x^3}{6} + \cdots \tag{4.46}$$

By choosing $x = \frac{1}{\beta A}$, it is possible to use only the two first terms of the series, which is valid, since when employing negative feedback one will generally aim to have $\beta \cdot A \gg 1$. In this case

$$\frac{v_o}{v_s} = \frac{1}{\beta} \left(1 - \frac{1}{\beta A} \right) \tag{4.47}$$

$$\frac{v_o}{v_s} = \left(1 + \frac{R_2}{R_1} \right) \times \left(1 - \frac{1}{\beta A} \right) \tag{4.48}$$

Operational Amplifiers

The realistic closed-loop gain becomes corrected by the term $\frac{1}{\beta A}$, which takes into account the OpAmp's finite open-loop gain. The following definitions are adopted:

$$G_{real} = \frac{v_o}{v_s} \qquad (4.49)$$

$$G_{ideal} = 1 + \frac{R_2}{R_1} \qquad (4.50)$$

It can be stated that the percentage error when considering the real closed-loop relative to the ideal closed-loop is given by

$$E\% = \left|\frac{G_{real} - G_{ideal}}{G_{ideal}}\right| \times 100\% = \frac{1}{|\beta A|} \times 100\% \qquad (4.51)$$

Inverter circuit

In order to determine the closed-loop gain for the inverter circuit seen in Figure 4.15, a procedure analogous to the previously described calculation is performed

$$I_1 = \frac{v_s - v_d}{R_1} \qquad (4.52)$$

$$I_2 = \frac{v_d - v_o}{R_2} \qquad (4.53)$$

$$v_o = -v_d \times A \qquad (4.54)$$

Calculating the value of v_d in terms of v_o and v_s, from (4.1) and (4.2), and substituting into (4.3), after some manipulation the following result is

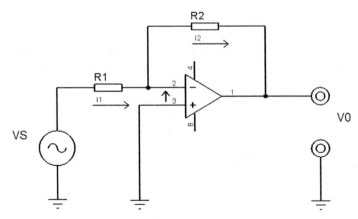

Figure 4.15 Inverter circuit.

obtained,

$$\frac{v_o}{v_s} = -\frac{R_2}{R_1 + R_2} \times \frac{A}{1 + \frac{R_1}{R_1+R_2} \times A}. \qquad (4.55)$$

Defining

$$\beta = \frac{R1}{(R1 + R2)}, \qquad (4.56)$$

then

$$\frac{v_o}{v_s} = -\frac{R_2}{R_1 + R_2} \times \frac{A}{1 + \beta A} \qquad (4.57)$$

$$\frac{v_o}{v_s} = -\frac{R_2}{R_1 + R_2} \times \frac{1}{\beta} \times \frac{1}{1 + \frac{1}{\beta A}} \qquad (4.58)$$

$$\frac{v_o}{v_s} = -\frac{R_2}{R_1} \times \frac{1}{1 + \frac{1}{\beta A}} \qquad (4.59)$$

$$\frac{v_o}{v_s} = -\frac{R_2}{R_1} \times \left(1 - \frac{1}{\beta A}\right) \qquad (4.60)$$

Joint Influence of Gain and Impedances

For the calculation of the closed-loop gain, the influence of the finite gain (A), the non-infinite input resistance ($R1$) and the non-zero output resistance (R_o) of the OpAmp on the inverter circuit shown in Figure 4.16 will be taken into account.

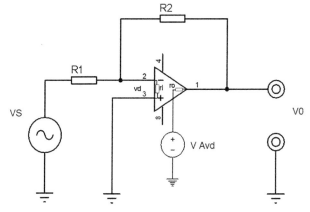

Figure 4.16 Inverter circuit showing the OpAmp's input and output resistances.

52 Operational Amplifiers

At the end of the process, it will be shown by numerical substitution of the usual values for the elements employed, that the OpAmp can be considered as ideal, with good precision, for low frequencies.

Firstly, it is possible to determine the v_o/v_d ratio in order to apply Miller's theorem to resistance $R2$:

$$K = \frac{v_o}{v_d} \qquad (4.61)$$

$$v_o = -v_d A - R_o i_o \qquad (4.62)$$

$$i_o = \frac{v_o - v_d}{R_2} \qquad (4.63)$$

Replacing the third equation into the second one yields:

$$v_o = -v_d A - R_o \frac{v_o - v_d}{R_2} \qquad (4.64)$$

From which the following result can be obtained

$$\frac{v_o}{v_d} = \frac{-A + R_o Y_2}{1 + R_o Y_2} = K, \text{ in which } Y_2 = \frac{1}{R_2} \qquad (4.65)$$

Applying Miller's theorem to resistance $R2$ and drawing the equivalent input circuit in terms of the admittances yields the circuit seen in Figure 4.17 from which the following equations are formulated:

$$i = v_s \times \frac{Y_1 - Y_{eg}}{Y_1 + Y_{eg}}, \text{ in which } Y_{eg} = Y_i + (1-k)Y \qquad (4.66)$$

$$v_d = i \times \frac{1}{Y_{eg}} \qquad (4.67)$$

$$v_d = \frac{Y_1}{Y_1 + Y_{eg}} \times v_s \qquad (4.68)$$

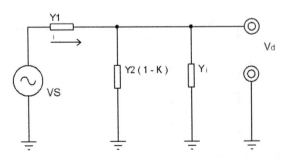

Figure 4.17 Equivalent input circuit in terms of admittances.

4.3 Real Operational Amplifier

Replacing this expression for v_d into (4.61) yields

$$v_o = K \times v_d = K \times \frac{Y_1}{Y_1 + Y_{eg}} \times v_s \qquad (4.69)$$

$$\frac{v_o}{v_s} = \frac{KY_1}{Y_1 + Y_i + Y_2 - KY_2}. \qquad (4.70)$$

Dividing the numerator and denominator of the second member of the latter expression by $-KY_2$ yields

$$\frac{v_o}{v_s} = -\frac{Y_1}{Y_2} \times \frac{1}{1 - \frac{(Y_i + Y_1 + Y_2)}{KY_2}} \qquad (4.71)$$

and substituting the value of K, obtained previously, into the expression results in

$$\frac{v_o}{v_s} = -\frac{R_2}{R_1} \times \frac{1}{1 + \frac{(Y_i + Y_1 + Y_2)(1 + R_o Y_2)}{Y_2(A - R_o Y_2)}} \qquad (4.72)$$

From obtained expression it can be concluded that increasing the output impedance (R_o), or decreasing the open-loop gain (A) and input impedance (R_i), will force the real closed-loop gain to be less than theoretical gain defined by the ratio $R2/R1$.

It is noteworthy that for the usual values for resistors $R1$ and $R2$ and for typical parameter values of an integrated OpAmp, the error incurred is quite small, as shown in the following example.

Example: Numerical amplification. Let the values for the circuit elements be given by:

$$R1 = R2 = 10 \text{ k}\Omega \qquad Y1 = Y2 = 10^{-4} \text{ S}$$
$$R_i = 500 \text{ k}\Omega \qquad Y_i = 2 \times 10^{-6} \text{ S}$$
$$R_o = 100 \text{ } \Omega \qquad A = 10^5$$

Replacing the given values into the closed-loop gain expression yields

$$\frac{v_o}{v_s} = -1 \times \frac{1}{1 + \frac{(2 \times 10^{-6} + 2 \times 10^{-4})(1 + 10^2 \times 10^{-4})}{10^{-4}(10^5 - 10^2 \times 10^{-4})}}$$

$$\frac{v_o}{v_s} = -\frac{1}{1 + \frac{220 \times 10^{-6} \times 1.01}{10}} = -\frac{1}{1 + 2 \times 10^{-5}}$$

Therefore, the error obtained when considering the gain equal to -1 is 0.002%, which is quite small.

4.3.2 Offset Voltage

When grounding the two inputs of an OpAmp, one would expect the output voltage (v_o) to be zero, however, this does not happen and an output voltage is observed, which may be saturated at either of the OpAmps supply rail voltages, $+V$ or $-V$, depending on the sample used.

To take into account the fact that the OpAmp is unbalanced, and to explain its real behavior, a model was built, in which the OpAmp is considered to be perfectly balanced and a continuous voltage source is connected to one of the inputs, as seen in Figure 4.18, thus justifying the non-zero output voltage when both inputs are grounded. This voltage source is called the offset (v_{os}), or residual, voltage.

A typical value, and sometimes the maximum and minimum values, of the offset voltage, are provided by the manufacturer in datasheets for any particular OpAmp type. Nonetheless, it is always possible to measure the polarity and value of the offset voltage through the use of the circuit seen in Figure 4.19, in which the values for $R1$ and $R2$ are such that:

(a) $R2 \gg R1$ in order to obtain a high closed-loop gain and therefore an easily measurable output voltage.
(b) $R1$ with a small resistance, so as to eliminate the influence of bias current on the output voltage.

The offset voltage becomes an important error element whenever the closed-loop gain is high, or when the OpAmp is operating as an integrator. In either case, its effect can be minimized by employing a compensation circuit, such as shown in Figure 4.20.

4.3.3 Bias Current

Although in the ideal OpAmp model it is said that no current flows into either of the input terminals, in reality, there is a small current that flows into both inputs, biasing the input transistors.

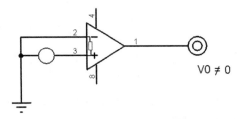

Figure 4.18 Model for the OpAmp including the off-set voltage.

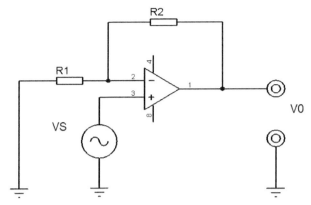

Figure 4.19 Setup to evaluate the offset voltage of the OpAmp, $R2 \gg R1$ and $R1$ have small resistance.

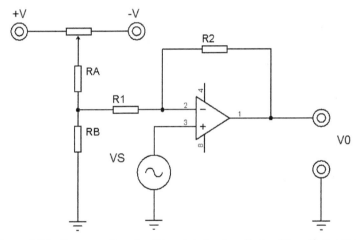

Figure 4.20 Compensating circuit to minimize the effects of the offset voltage.

In case the differential input stage consists of bipolar transistors, the bias current corresponds to the base current, while if they are field-effect transistors (FETs), it corresponds to a leakage current. For the latter, the current observed will evidently be much smaller. The currents flowing through the inputs (+) and (−) will be designated, respectively, as I^+ and I^-.

Customarily, the value of the bias current is defined as the arithmetic mean between the two input currents:

$$I_b = \frac{I^+ + I^-}{2} \quad \text{(bias current)}. \tag{4.73}$$

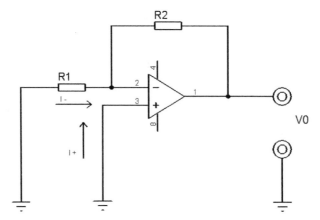

Figure 4.21 Representation of bias currents.

The bias currents are interchangeably denominated I^+, I^- and I_b, although I_b is the only value provided in OpAmp datasheets.

Another important element is the offset current, defined as the difference between currents I^+ and I^-

$$I_{os} = |I^+ - I^-|. \qquad (4.74)$$

Influence of currents I^+ and I^-

Consider Figure 4.21 and suppose at first that v_o can be zero. Under this condition, voltage v^- will be slightly negative since it will be given by $-(R1//R2) \times I^-$ whereas $v^+ = 0$, therefore v^+ is different from v^- and the output voltage v_o will be nonzero. The bias currents I^+ and I^- are, in this case, will cause an error in the operation of the circuit.

One way to eliminate, or at least minimize, the error produced by currents I^+ and I^- is to interleave a resistor of value $R3 = R1//R2$ at input (+). In this case, one can initially assume that $v_o = 0$ and prove that it is possible. Since $v_o = 0$, input voltage (−) will be given by $v^- = -(R1//R2) \times I^-$ and the input voltage (+) will be $v^+ = -R3 \cdot I^+$. If $I^+ = I^-$ and $R3 = (R1//R2)$, then $v^+ = v^-$, consequently the output voltage v_o will be zero, Figure 4.22.

Influence of the offset current

Once the value of $R3$ (Figure 4.22) is equal to the parallel $R1//R2$, the compensation of currents I^+ and I^- will only be perfect if $I^+ = I^-$,

4.3 Real Operational Amplifier   57

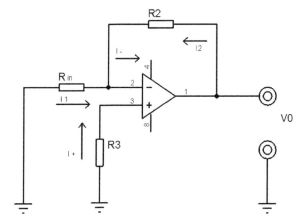

Figure 4.22 Circuit that minimizes the error produced by the polarization currents.

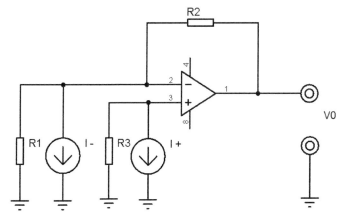

Figure 4.23 Circuit that has the input currents represented by current sources.

otherwise, a small error will still exist, and it will be a function of the offset current and of resistance R2, as shown below.

Input currents (−) and (+) can be represented by two current sources I^- and I^+, as seen in Figure 4.23, in which the OpAmp is considered to be ideal.

Applying Thevenin's theorem, the circuit can be modified to the configuration seen in Figure 4.24, in which two voltage sources appear, one in the inverting input, with value $-R1 \cdot I^-$, and the other in the non-inverting input, with value $-R3 \cdot I^+$. Because the system is linear, the superposition principle can be applied and the output voltage v_o will be the sum of two voltages, v_{o1}

58 Operational Amplifiers

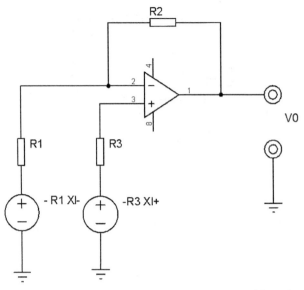

Figure 4.24 Circuit in which the input currents are represented by voltage sources.

and v_{o2}, in which

$$v_{o1} = -R_3 \times I^+ \times \left(1 + \frac{R_2}{R_1}\right) \quad (4.75)$$

$$v_{o2} = -\frac{R_2}{R_1} \times (-R_1 I^-) = +R_2 I^- \quad (4.76)$$

Therefore,

$$v_o = v_{o1} + v_{o2} \quad (4.77)$$

$$v_o = -R_3 \times I^+ \times \left(1 + \frac{R_2}{R_1}\right) - \frac{R_2}{R_1} \times (-R_1 I^-) \quad (4.78)$$

$$v_o = -I^+ \times R_3 \times \frac{R_1 + R_2}{R_1} + R_2 I^- \quad (4.79)$$

$$v_o = -I^+ \times R_2 + I^- \times R_2 \quad (4.80)$$

$$v_o = R_2(I^- - I^+) \quad (4.81)$$

Recalling that the definition given to the offset current is

$$v = \pm R_2 \times I_{os}. \quad (4.82)$$

4.3.4 Influence of temperature

The offset voltage and offset current are temperature dependent, and therefore the output voltage of a feedback OpAmp will have an error that will vary with temperature. Thus, if the circuit indicated in Figure 4.22 is being used in an environment where temperature is variable, the output voltage will evolve according to the expression

$$\frac{dv_o}{dT} = \pm \left(1 + \frac{R_2}{R_1}\right) \times \frac{dv_{os}}{dT} \pm R_2 \times \frac{dI_{os}}{dT}, \tag{4.83}$$

in which $\frac{dv_{os}}{dT}$ and $\frac{dI_{os}}{dT}$ are the offset voltage and offset current derivatives, respectively. These two terms allow for the OpAmp's evaluation in terms of temperature performance, according to the circuit.

4.3.5 Common-mode rejection ratio

Usually, a differential amplifier should provide an output voltage that is only proportional to the difference between the two input signals, however, it is found in its actual operation that the output also depends on the average value or the arithmetic average of the signals applied to the two inputs, this value is called common-mode voltage.

Consider the differential amplifier shown in Figure 4.25, in which $A1$ is the gain relative to input 1 and $A2$ is the gain relative to input 2. Then

$$A_1 = \frac{v_o}{v_1} \quad \text{with } v_2 = 0, \tag{4.84}$$

$$A_2 = \frac{v_o}{v_1} \quad \text{with } v_1 = 0, \tag{4.85}$$

and

$$v_o = A_1 v_1 + A_2 v_2., \tag{4.86}$$

Consider the following definition

$$v_d = v_1 - v_2, \tag{4.87}$$

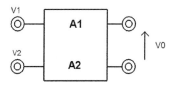

Figure 4.25 Quadripole representation of the operational amplifier.

and
$$v_c = \frac{v_1 + v_2}{2}, \tag{4.88}$$

From which it can be derived that
$$v_1 = \frac{1}{2}v_d + v_c, \tag{4.89}$$

and
$$v_2 = -\frac{1}{2}v_d + v_c., \tag{4.90}$$

Replacing these results into the expression for the output voltage, and grouping in terms of v_d and v_c yields
$$v_o = \frac{1}{2}(A_1 - A_2)v_d + (A_1 + A_2)v_c \tag{4.91}$$

The coefficients multiplying v_d and v_c are called, respectively, the differential gain (A_d) and common mode gain (A_c):
$$A_d = \frac{1}{2}(A_1 - A_2), \tag{4.92}$$
$$A_c = A_1 + A_2. \tag{4.93}$$

Then
$$v_o = A_d v_d + A_c v_c, \tag{4.94}$$

or yet
$$v_o = A_d\left(v_d + \frac{A_c}{A_d} \times v_c\right) \tag{4.95}$$
$$v = A_d\left(v_d + \frac{v_c}{A_d/A_c}\right). \tag{4.96}$$

Defining $\rho = \frac{A_d}{A_c}$, then
$$v_o = A_d\left(v_d + \frac{v_c}{\rho}\right), \tag{4.97}$$

in which the ratio $\frac{A_d}{A_c} = \rho$ is called the common-mode rejection ratio (CMRR) and represents a merit figure for the differential amplifier. It can be observed that the higher the value of ρ, the smaller the influence of the common-mode voltage.

In closed-loop configurations using an OpAmp as an active element, the common-mode voltage is only considered in non-inverter circuits, as for inverter circuits its value is zero, Figure 4.26.

4.3 Real Operational Amplifier

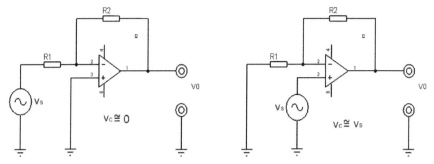

Figure 4.26 Common-mode voltage, for a circuit with negative feedback: (a) Input voltage applied to the inverting input; (b) Input voltage applied to the non-inverting input.

4.3.6 Frequency Response

The OpAmp's open-loop gain varies greatly with frequency, its amplitude and phase decrease as the operating frequency increases, so even with the use of negative feedback, the amplifier can be led to instability if certain conditions are not met.

The behavior of the closed-loop amplifier is studied, assuming that its open-loop gain can have one, two, three or more poles.

When the open-loop gain of an OpAmp has only one cutoff frequency or one pole, its transfer function has the following derivation.

OpAmp considered as a first-order system

When the open-loop gain of an OpAmp has only one cutoff frequency or one pole, its transfer function has the following form,

$$A(\omega) = \frac{A_o}{1 + j\frac{\omega}{\omega_c}} = \frac{A_o}{1 + j\frac{f}{f_c}} \qquad (4.98)$$

in which f_c is the cutoff frequency, defined as

$$|A(\omega_c)| = \frac{A_o}{\sqrt{2}} \qquad (4.99)$$

Considering the gain amplitude, in dB

$$|A(\omega_c)|\text{dB} = 20\log|A(\omega_c)| = 20\log A_o - 20\log 2^{\frac{1}{2}} \qquad (4.100)$$

$$|A(\omega_c)|\text{dB} = |A_o|\text{dB} - 3\text{ dB}, \qquad (4.101)$$

that is why ω_c, or f_c, is the frequency called the -3 dB cutoff frequency.

62 Operational Amplifiers

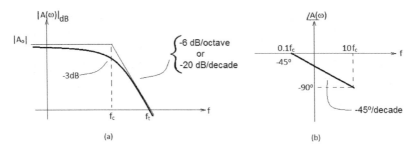

Figure 4.27 Bode diagrams: (a) amplitude; (b) phase.

The phase function, $\theta(\omega)$, or $\angle A(\omega)$, as a function of frequency is given by,

$$\theta(\omega) = -\arctan \frac{\omega}{\omega_c} \qquad (4.102)$$

The Bode amplitude and phase curves are shown in Figure 4.27

The frequency f_t for which the gain amplitude is equal to unit is called the transition frequency or unit gain frequency, f_o, its value is determined as follows:

$$|A(\omega)| = \frac{A_o}{\sqrt{1 + \left(\frac{f}{f_c}\right)^2}} \qquad (4.103)$$

$$1 + \left(\frac{f_t}{f_c}\right)^2 = A_o^2 \qquad (4.104)$$

$$\left(\frac{f_t}{f_c}\right)^2 = A_o^2 - 1 \cong A_o^2 \qquad (4.105)$$

$$f_t = A_o \times f_c \qquad (4.106)$$

The behavior of the OpAmp in the non-inverting circuit of Figure 4.28 is verified in the following. The closed-loop gain $A_{CL}(\omega)$ is given by

$$A_{CL}(\omega) = \frac{A(\omega)}{1 + \beta \cdot A(\omega)} \qquad (4.107)$$

in which $\beta = \frac{R_1}{R_1+R_2}$ is the feedback rate.

$$A_{CL}(\omega) = \frac{\frac{A_o}{1+j\omega/j_c}}{1 + \beta \times \frac{A_o}{1+j\omega/j_c}} \qquad (4.108)$$

$$A_{CL}(\omega) = \frac{A_o}{1+\beta A_o} \times \frac{1}{1+j\frac{\omega}{(1+\beta A_o)\omega_c}} \qquad (4.109)$$

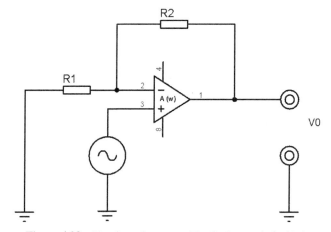

Figure 4.28 Non-inverting setup. The OpAmp gain is $A(w)$.

or

$$A_{CL}(\omega) = \frac{A_{CL}(0)}{1 + j\frac{\omega}{\omega_c(1+\beta A_o)}} \therefore A_{CL}(\omega) = \frac{A_{CL}(0)}{1 + j\frac{\omega}{\omega'_c}} \quad (4.110)$$

in which $\omega'_c = (1 + \beta A_o)\omega_c$.

The closed-loop gain was reduced at low frequencies by the factor $(1 + \beta A_o)$, however, the -3 dB cutoff frequency increased to $(1 + \beta A_o) \times f_c$. The product of the low frequency gain by the -3 dB cutoff frequency remained constant,

$$A_o \times \omega_c = A_{CL}(0) \times \omega'_c \quad (4.111)$$

that is

$$A_o \times \omega_c = \frac{A_o}{1 + \beta A_o} \times (1 + \beta A_o) \times \omega_c \quad (4.112)$$

The Bode diagram of the closed-loop gain is shown in Figure 4.29.

$$A_{CL}(0) = \frac{A_o}{(1 + \beta A_o)} \cong \frac{1}{\beta} \quad (4.113)$$

$$f_1 = (1 + \beta A_o) \times f_c \quad (4.114)$$

The graphical operation in dB is simplified since,

$$A_{CL}(\omega)dB = 20\log\frac{A(\omega)}{1 + \beta A(\omega)} \quad (4.114)$$

Considering that $\beta A(\omega) \gg 1$, one obtains

$$A_{CL}(\omega)dB = 20\log A(\omega) - 20\log \beta A(\omega) \quad (4.115)$$

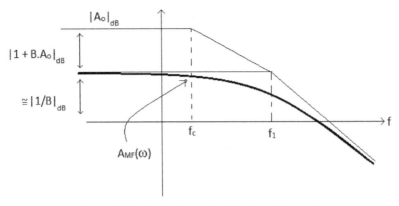

Figure 4.29 Bode diagram for the closed-loop gain.

or
$$A_{CL}(\omega)dB = A(\omega)dB - \beta A(\omega)dB \tag{4.116}$$

in which $A(\omega)$ is the open-loop gain, in dB, and $\beta A(\omega)$ is the closed-loop gain, also in dB.

Therefore, the closed-loop gain can be obtained from the graphical difference between the open- and closed-loop gains, measured in dB; which can be obtained from the technical specifications and from the circuit design, respectively.

Loop gain plays an important role in checking the amplifier's stability at high frequencies.

In low frequencies considering that $\beta A_o \gg 1$, therefore

$$|A_{CL}(0)|dB = \left|\frac{A_o}{1+\beta A_o}\right|dB \cong \left|\frac{1}{\beta}\right|dB \tag{4.117}$$

Stability: Nyquist Criterion

In a simplified form, Nyquist's criterion states that if the curve for the closed-loop gain $\beta \times A(\omega)$, in polar coordinates, involves the point -1, the system is considered unstable.

More clearly, a system is said to be stable if

$$\angle \beta A(\omega) = -180° \tag{4.118}$$

and
$$|\beta A(\omega)| < 1. \tag{4.119}$$

And is unstable if
$$\angle \beta A(\omega) = -180° \qquad (4.120)$$
and
$$|\beta A(\omega)| \geq 1. \qquad (4.121)$$

From the closed-loop gain expression, $A_{CL}(\omega)$, it can be seen that if $\beta \times A(\omega) = 1\angle -180°$, the amplifier will oscillate. Since in this case, β is constant and the maximum phase-shift introduced by $A(\omega)$ is $-90°$, the circuit will be stable for whatever the value is given to β.

5
Circuits with Operational Amplifiers

This chapter describes several circuits with OpAmps that can be used for various applications, some of the transfer function expressions have been obtained in previous chapters.

Operational amplifiers have a very interesting feature that when in a closed negative feedback loop, the input voltages of the positive and negative terminals are virtually equal. This allows for multiple applications as shown in this chapter (Millman and Halkias, 1972).

Thus, the basic rules when dealing with feedback loop OpAmps circuits are:

1. Voltages at the OpAmp input terminals tend to be equal,
$$v^+ = v^- \tag{5.1}$$

2. OpAmp input currents are theoretically zero because the input resistance is very high,
$$I^+ = I^- = 0 \tag{5.2}$$

3. The amplification factor is theoretically infinite. However, in practice the output voltage is limited by the OpAmp's supply voltages:
$$A \to \infty; \max(V_o) = V^+; \quad \min(V_o) = V^- \tag{5.3}$$

With these basic rules, it is possible to obtain several interesting applications for OpAmp circuits.

5.1 Inverting Amplifier

The most widely used constant gain amplifier is the inverter amplifier. As described previously, this circuit is intended to amplify the input signal while shifting the phase by $180°$.

68 Circuits with Operational Amplifiers

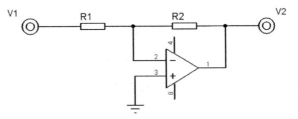

Figure 5.1 Inverting amplifier.

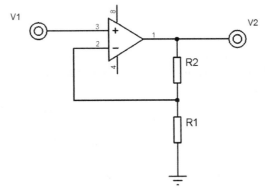

Figure 5.2 Non-inverting amplifier.

Input resistance,
$$R_i = R_1 \tag{5.4}$$

The output resistance of the circuit is low (theoretically the output of the OpAmp itself).

Gain,
$$\frac{v_2}{v_1} = -\frac{R_2}{R_1} \tag{5.5}$$

5.2 Non-inverting Amplifier

This circuit amplifies the signal without inverting it, however, it tends to be more unstable than the inverting amplifier.

Input resistance has high value (theoretically the input of the OpAmp itself). The output resistance of the circuit is low. The gain is

$$\frac{v_2}{v_1} = 1 + \frac{R_2}{R_1} \tag{5.6}$$

5.3 Oscillators

Oscillators are circuits that allow a periodic oscillating signal to be created without an input signal. It is usually based on an amplifier circuit and a positive feedback loop, which induces operation instability, resulting in oscillation.

In order for the oscillator circuit to function properly, two conditions must be met; these are called the Barkhausen conditions:

1. The closed-loop gain must be 1.
2. The total phase shift of the signal in the circuit must be $0°$ or a multiple of $360°$.

Oscillators have several applications, such as, in clocks, audio synthesizers, radiofrequency systems and in the control of electrical machines.

Two types of oscillators are discussed: the RC phase shift oscillator and the Wien bridge oscillator. However, the reader is encouraged to investigate the operation and construction of other types, such as the Hartley oscillator, the crystal oscillator, the relaxation oscillator, and the astable oscillator.

5.3.1 RC Phase Shift Oscillator

With negative feedback, the amplification stage produces a $180°$ phase shift between input and output. So if an RC stage is added, it produces a complementary phase shift of another $180°$, yielding a $360°$ shift, equivalent to $0°$, thus it provides positive feedback. This can be done with the circuit seen in Figure 5.3, in which the RC stages create a $180°$ degree offset.

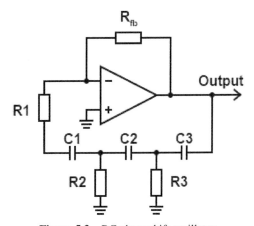

Figure 5.3 RC phase shift oscillator.

70   Circuits with Operational Amplifiers

Question: Why are at least three RC stages required to result in a $180°$ phase shift?

For $R_1 = R_2 = R_3 = R$ and $C_1 = C_2 = C_3 = C$. The phase shift for each section is $60°$ and the oscillation frequency for the resulting circuit is

$$f_r = \frac{1}{2\pi RC \sqrt{2N}} \qquad (5.7)$$

in which N is the number of RC stages used. In the example, three stages are used.

Question: Calculate the parameters of a 3-stage oscillator with a frequency of 6.5 kHz. Use 1 nF capacitors. (Answer: $R \approx 10$ kΩ and $RF \approx 290$ kΩ).

5.3.2 Wien Bridge Oscillator

The Wien Bridge Oscillator is preferred for use in commercial audio generators as it can reach frequencies as low as 5 Hz and up to 1 MHz, using a smaller quantity of passive elements. The circuit is named after the frequency-selective form of the Wheatstone bridge, being a two-stage RC circuit with good resonant frequency stability, low distortion, and relatively easy tuning.

The oscillation frequency of the Wien bridge is

$$f_r = \frac{1}{2\pi RC} \qquad (5.8)$$

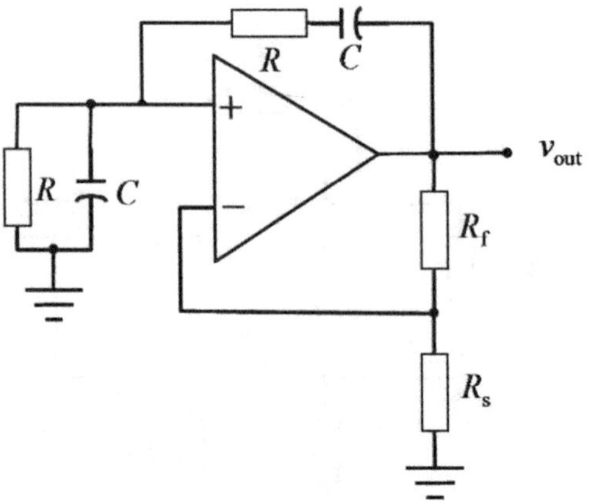

Figure 5.4   Wien bridge oscillator.

For sinusoidal oscillations to begin, the voltage gain given by the non-inverting amplifier, with R_f and R_s, must be greater than or equal to 3, thus

$$\frac{R_f}{R_s} + 1 \geq 3 \quad (5.9)$$

$$\therefore$$

$$R_f \geq 2R_s \quad (5.10)$$

An interesting detail is that it used to be common to put an incandescent lamp in the place of resistor R_s, because it allows a large attenuation of signal distortion if there are variations in the supply voltage or in the circuit temperature. A problem related to using a lamp is that it could also add distortion if it experienced vibration.

Question: Calculate the parameters of a Wien bridge oscillator that generates a 5.2 kHz sinusoidal signal. Consider a capacitor with $C = 3$ nF.
(Answer: $R = 10.2$ kΩ; $R_f = 100$ kΩ and $R_s = 50$ kΩ)
Determine the maximum and minimum oscillator frequencies if the 3 nF capacitor is now replaced by a variable capacitor from 1 nF to 1000 nF.

5.4 Buffer

The buffer circuit is composed only of an OpAmp with negative feedback; it has unit gain and has no phase inversion (Alley and Atwood, 1973).

The input resistance is very high and the output resistance is very low. This circuit is used when one desires to copy the input onto the output; it is applied in impedance matching networks, isolation between different circuit stages, and current boosters.

When the buffer receives a signal through a series resistor, it is common to place a resistor of the same resistance in the feedback loop, to balance the

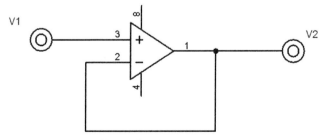

Figure 5.5 Buffer circuit.

gain and keep it equal to one. This technique is used in matching impedances between a signal generator and an amplifier with small input impedance, such as an LNA.

The gain is given by

$$\frac{v_2}{v_1} = 1 \qquad (5.11)$$

5.5 Comparator

The comparator simply consists of a high gain amplifier connected to two input voltages, or an input voltage and a reference voltage.

Voltage comparators have a wide range of applications in controllers, alarms, sensors, level detectors, switches, etc. They can be found in ICs dedicated to this function, such as LM339, LM239, and LM139.

The input resistance has typical values between 10 kΩ and 1 MΩ. The output resistance has typical values between 1 kΩ and 10 kΩ. The gain is given by the open-loop gain, that is

$$v_o = A(v_+ - v_-) \qquad (5.12)$$

Since the gain is very high, and the voltage saturates at the supply voltage values, then, in reality, the output is given by:

$$v_o = \begin{cases} V^+, & \text{para } v_+ > v_- \\ V^-, & \text{para } v_+ < v_- \end{cases} \qquad (5.13)$$

If one wishes to eliminate the negative voltage when the input is lower than the reference value, one can add a diode so that there will only be a positive output.

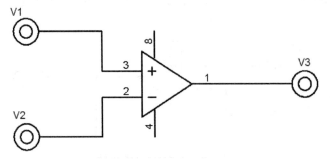

Figure 5.6 Voltage comparator circuit.

5.6 Adder 73

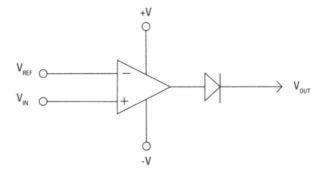

Figure 5.7 Comparator circuit with positive output only.

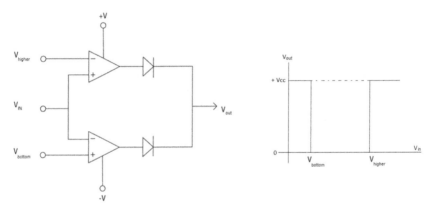

Figure 5.8 Window comparator and it is correspondent output *versus* input graph.

A useful configuration that can be obtained using two comparators is the window comparator, which will only provide a non-zero output when the input voltage is outside a certain range, below the set inferior reference or above the upper reference.

5.6 Adder

This circuit makes it possible to add two time-varying voltages, it is a variation of the inverting amplifier circuit.

The circuit's input resistance is

$$R_i = R \qquad (5.14)$$

74 Circuits with Operational Amplifiers

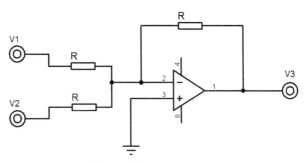

Figure 5.9 Adder circuit.

The circuit's output resistance is
The gain is

$$v_3 = -(v_1 + v_2) \tag{5.15}$$

One of the most interesting applications of the adder is the construction of a digital-to-analog converter. In fact, if the k signal sources, b_k, are assumed to be 1 V or 0 V depending on the logical value of the bits of a digital word, and the corresponding resistors R_k are given values according to the bit's position inside the word, for example:

$$R_1 = R, R_2 = R/2, R_3 = R/4 \ldots R_k = R/2^{k-1}, \tag{5.16}$$

then, the voltage expression at the OpAmp output is

$$v_o = (2^{k-1}b_k + \cdots + 8b_4 + 4b_3 + 2b_2 + b_1) \tag{5.17}$$

For example, the digital words 10011 and 00001 (in decimal: 19 and 1, respectively) lead to output voltage values

$$v_o = -(16 + 0 + 0 + 2 + 1) = -19 \text{ V}$$

and

$$v_o = -(0 + 0 + 0 + 0 + 1) = -1 \text{ V}$$

respectively. Of course, the value of resistor R can be scaled to redefine the amplitude range of the output voltage.

5.7 Subtractor

This circuit is the combination of inverting and non-inverting amplifiers, so the output may be the amplification of the difference between two time-varying signals.

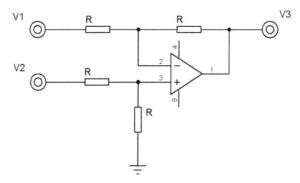

Figure 5.10 Subtractor circuit, or differential amplifier.

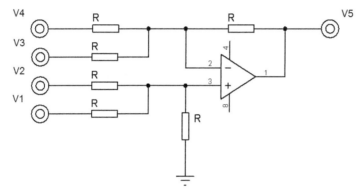

Figure 5.11 Adder/subtractor circuit.

The input resistance is variable with the voltage, as there is no virtual ground. The output resistance of the circuit is small.

The gain is given by

$$v_3 = (v_2 - v_1) \qquad (5.18)$$

5.8 Adder/Subtractor

This circuit, in turn, is the combination of the adder with the subtractor, so that it can simultaneously perform addition and subtraction of several time-varying signals (Silva, 2013).

The input resistance is variable with voltage as there is no virtual ground. The output resistance of the circuit is small. The gain is

$$v_5 = (v_1 + v_2) - (v_3 + v_4) \qquad (5.19)$$

76 Circuits with Operational Amplifiers

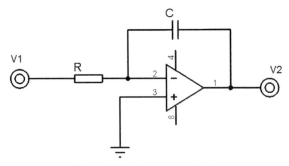

Figure 5.12 Integrator circuit.

5.9 Integrator

With the use of a capacitor, it is possible to transform the inverter amplifier into a circuit whose output voltage is the integral of the input voltage. The circuit seen in Figure 5.12 can be used for this purpose. The demonstration is in the following.

1. $v^+ = v^- = 0$ (5.20)
2. $I_1 = \dfrac{v_s}{R}$ (5.21)
3. $I_c = I_1$ (5.22)
4. $Q_c = \displaystyle\int I_c \, dt$ (5.23a)

$$Q_c = \int \dfrac{v_s}{R} dt \qquad (5.23b)$$

$$v_o = -\dfrac{Q_c}{C} = -\dfrac{1}{RC}\int v_s \, dt \qquad (5.24)$$

The input resistance is R. The circuit's output resistance is small. The gain is given by

$$v_2 = -\dfrac{1}{RC}\int_{t_o}^{t} v_1 \, dt \qquad (5.25)$$

5.10 Differentiator

Similarly, it is possible to use the inverting amplifier circuit to build a differentiator, which yields the input signal's derivative.

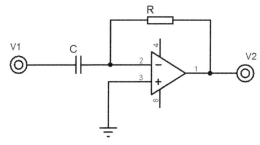

Figure 5.13 Differentiator circuit.

The input resistance is:

$$R_i = \frac{1}{j\omega C} \tag{5.26}$$

Note that the input resistance varies with frequency: for lower frequencies the resistance increases, for higher voltages it decreases (Mayaram, 2008). The output resistance of the circuit is small, and the gain is

$$v_2 = -RC\frac{dv_1}{dt} \tag{5.27}$$

5.11 Instrumentation Amplifier

The instrumentation amplifier, also known as a widely used alternative for instrumentation applications, as it has a steady high input resistance, while a subtractor would have a variable resistance. Also, it has a high common-mode rejection ratio. This circuit can be used in medical applications, such as in neuronal signal sensors.

The instrumentation amplifier gains are given by

$$v_5 = (v_1 + v_2) - (v_3 + v_4) \tag{5.28}$$

Note that this circuit uses a non-inverting amplifier as a buffer for each input.

5.12 Shifter

The shifter is a circuit that does not necessarily present an amplitude gain; it is essentially used to shift the phase of the input signal, as shown in Figure 5.15.

Note that the gain is always unitary and that the phase varies from $0°$ to $180°$, according to the value or the resistor, r, which can be set as a variable resistance of a potentiometer.

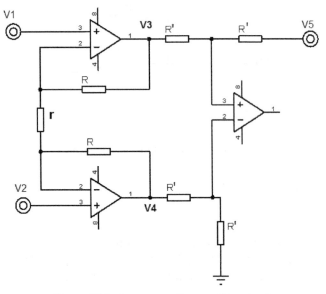

Figure 5.14 Instrumentation amplifier signal.

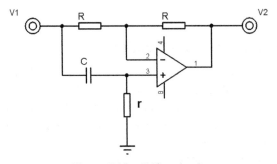

Figure 5.15 Shifter circuit.

The circuit's output resistance is small, and the gain is given by

$$v_2 = -\left(\frac{1 - j\omega Cr}{1 + j\omega Cr}\right) \cdot v_1 \qquad (5.29)$$

5.13 Transresistance Amplifiers

This circuit (Figure 5.16) can be used as a current/voltage converter. The input resistance of the circuit is

$$R_i = 0. \qquad (5.30)$$

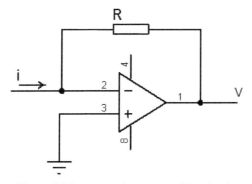

Figure 5.16 Transresistance amplifier circuit.

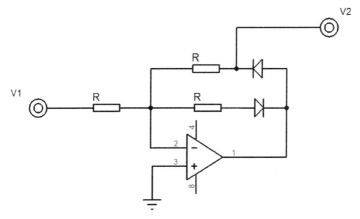

Figure 5.17 Precision rectifier circuit.

The circuit's output resistance is small, and the gain is

$$v = -i \cdot R \qquad (5.31)$$

5.14 Precision Rectifier

This circuit is a half-wave rectifier that rectifies signals that are below the diode voltage. The circuit's output resistance is small. The input resistance is

$$R_i = R \qquad (5.32)$$

Ideally, a diode should fully conduct when any positive is set as its input, and have a zero output for any negative input, however small. A real silicon

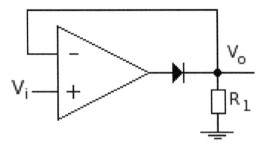

Figure 5.18 Alternative circuit for a half-wave precision rectifier.

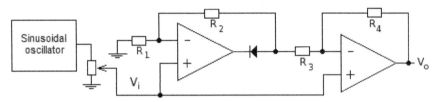

Figure 5.19 Full-wave precision rectifier.

diode, nonetheless, begins to conduct only once the direct voltage reaches a threshold of about 0.7 V. This can cause everything from unwanted distortion to complete loss of small signals.

There are other circuits that can be used for this same purpose, such as the circuit in Figure 5.18, also a half-wave precision rectifier.

Using an operational amplifier, the effect of the diode's direct conduction voltage can be minimized by placing it within the feedback loop.

There is also the full-wave rectifier circuit (also called the absolute value circuit), for which two OpAmps must be used.

The configuration seen in Figure 5.19 has necessarily a gain greater than one. The values of R_1, R_2, R_3 and R_4 must be adjusted so that the gain for negative signals is equal, but with a sign opposite to that of positive signals.

For example, to obtain gain equal to 2, one may use: $R_1 = R_2 = R_3 = R$ and $R_4 = 3 \cdot R$. Recommended values: $R_1 = R_2 = R_3 = 3,3 \text{ k}\Omega$ and $R_4 = 10 \text{ k}\Omega$.

5.15 Logarithmic Amplifiers

This circuit takes advantage of the logarithmic relationship between current and voltage in the transistor.

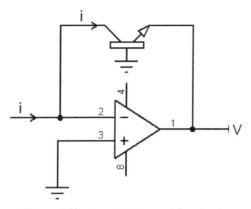

Figure 5.20 Logarithmic amplifier circuit.

The output resistance of the circuit is low and the input resistance is zero. However, it is common to have a resistor at the circuit's input, so that the current is given by $i = \frac{V}{R}$. The gain is

$$v = -a \cdot \ln b \cdot i = -a \cdot \ln b \cdot \left(\frac{V}{R}\right) \tag{5.33}$$

The circuit typically operates in 9 decades, from 10^{-3} to 10^{-12} A and, with it, it is possible to multiply and divide time-varying voltage signals, as well as perform exponential with the use of adders and subtractors.

Special amplifiers are manufactured for this purpose, called LogAmps, some ICs are AD640, AD641, ICL8048, LOG100 and 4127, and cost around 10 times the value of the components needed to build a discrete logarithmic amplifier. Normally, any LogAmp has temperature-dependent variations when operating, and sometimes it is necessary to compensate for this dependence.

Example: Show how one can multiply voltage signals and design the circuit that can do it. How is the circuit built for the anti-logarithmic function?
Answer:

5.16 High-impedance Differential Amplifier

Like the differential instrumentation amplifier, the circuit shown in Figure 5.22 has a high input impedance, yet uses one less amplifier.

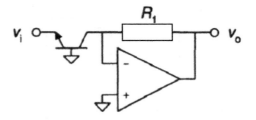

Figure 5.21 Anti-logarithmic circuit.

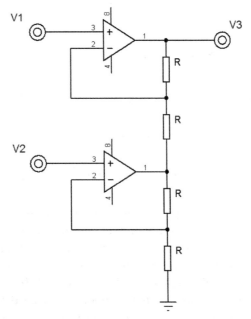

Figure 5.22 High-impedance differential amplifier circuit.

The gain is given by
$$v_3 = 2(v_1 - v_2) \tag{5.34}$$

5.17 Gyrator

The gyrator circuit allows simulating an impedance Z_{in}, which is a feature widely used to replace inductors in integrated circuits, since inductors have nonlinear characteristics and tend to produce harmonics, and are inevitably large to use inside embedded circuits.

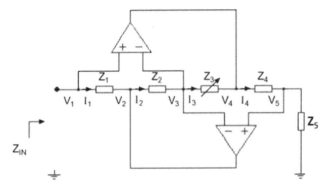

Figure 5.23 Gyrator circuit.

The impedance seen at the terminals shown in Figure 5.22 is given by

$$Z_{in} = \frac{Z_1 Z_3 Z_5}{Z_2 Z_4} \tag{5.35}$$

Note that, if Z_4 is replaced by a capacitor ($X_c = \frac{1}{j\omega C}$), and using resistors R in place of the other impedances, one has

$$Z_{in} = j\omega C R^2 = sCR^2 \tag{5.36}$$

which is equivalent to an inductor $L = CR^2$.

There is a device called a super inductor whose impedance is a function of the square of the frequency, that is

$$Z_{in} = s^2 L \tag{5.37}$$

6

Active Filters

The goal of this chapter is to introduce frequency filter concepts, instruct the student to understand and distinguish filters based on their frequency response and teach how to build active 1st-order and 2nd-order filters using generic circuit topologies.

Filters can be built using only passive components when so, they are called passive filters. However, the use of OpAmps guarantees some advantages over passive filters, as follows (Sedra and Smith, 2004).

- Only resistors and capacitors are required, eliminating the use of inductors, which are nonlinear and can distort the signal;
- They can have gain greater than 1 and rarely have losses, unlike passive filters;
- They are easily tuned;
- Low-frequency filters can be obtained with inexpensive components;
- They have low output impedances, allowing multiple filters to be connected in series.

Filters are linear circuits designed to pass or block signals at certain frequencies. That is, it has variable frequency gain, being classified according to their frequency response, such as Low-Pass, High-Pass, Band-Pass, and Band-Stop (Smith, 2004). Below, each classification is described and their frequency-gain graphs are shown.

Low-Pass Filter (LPF) – Filters high frequencies so that rapid variations in signal are attenuated. Theoretically, for frequencies below the filter resonant frequency, f_c, the output voltage is equal to the input voltage.

High-Pass Filter (HPF) – Low frequencies are filtered so that slow signal variations are mitigated. In theory, for frequencies above the filter resonant frequency, f_c, the output voltage is equal to the input voltage.

86 *Active Filters*

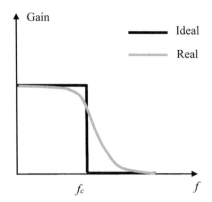

Figure 6.1 Frequency response of the LPF.

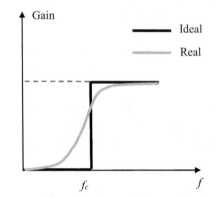

Figure 6.2 Frequency response of the HPF.

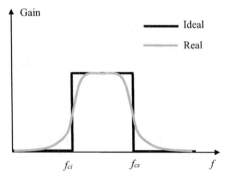

Figure 6.3 Frequency response of the BPF.

6.1 First-order filters

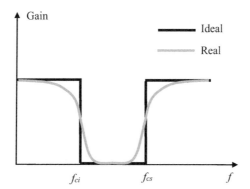

Figure 6.4 Frequency response of the BSF.

Band-Pass Filter (BPF) – Attenuates the signal for frequencies outside a given frequency range, between f_{ci} and f_{cs}.

Band-Stop Filter (BSF) – The signal is attenuated only for a defined frequency range, comprised between f_{ci} and f_{cs}. This type of filter is also known as a Band-Reject Filter and, when it has a narrow stopband, it can be called a Notch Filter.

The main parameters of a filter are (Kaufman, 1984):

- Passband – frequency range over which the signal suffers minimal attenuation;
- Stopband – Frequency range in which signals are greatly attenuated;
- Transition band – Frequency range between the passband and stopband, in which signals may have variable attenuation;
- Frequency Response ($H(\omega) = V_o/V_i$) – Function that determines the value of the gain as a function of frequency, filters can be of 1st order, 2nd order or n-th, according to their frequency responses;
- Cutoff Frequency (f_c, or ω_c) – Frequency for which the gain can be considered satisfactorily attenuating;
- 3 dB Bandwidth (Δf) – The range of frequencies in which the filter gain lies above -3 dB. An attenuation of 3 dB corresponds to half the maximum power. Or, in terms of Voltage amplitude, when the signal is at $\frac{1}{\sqrt{2}}$ of its maximum amplitude;
- Filter Quality Factor (Q) – Determines the format of the filter frequency response, it is defined as $Q \triangleq \frac{f_c}{\Delta f}$, in which f_c corresponds to the cutoff frequency and Δf is the -3 dB bandwidth.

88 Active Filters

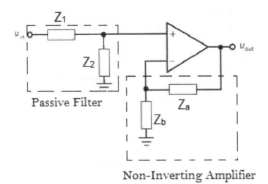

Figure 6.5 General topology of 1st-order filters.

6.1 First-order filters

Typically, 1st order filters tend to have a slower frequency response, with simpler functions. For 1st order filters, we will restrict ourselves to varying the passive elements of an RC passive filter, associated with the use of a non-inverting amplifier to increase the gain (Sedra and Smith, 2004). Consider the topology shown in Figure 6.5.

The passive filter part is basically a voltage divider, wherein, as the impedance Z_2 is much higher than Z_1, the output voltage tends to be equal to the input. And as Z_2 is much smaller than Z_1, the output voltage tends to be zero. When using capacitors, whose impedance is a function of frequency ($Z_c = \frac{1}{j\omega C}$), this section becomes a passive filter.

For the non-inverting amplifier section, only resistors are used, resulting in the previously derived gain expression

$$\frac{V_{out}}{V_{in}} = 1 + \frac{R_a}{R_b} \tag{6.1}$$

6.1.1 Low-pass Filter

For the low-pass filter in Figure 6.6, notice that as the frequency increases, the capacitor impedance decreases, so that for high frequencies the circuit output voltage tends to zero.

It can be shown that the frequency response expression of this circuit is given by (Padilha, 1993)

$$\frac{V_{out}}{V_{in}} = \frac{1}{1 + j\omega R_3 C_1} \cdot \left(1 + \frac{R_2}{R_1}\right) \tag{6.2}$$

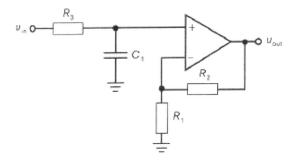

Figure 6.6 First-order LPF circuit.

and that the modulus of the expression, given by the numerator modulus divided by the denominator modulus) is

$$\left|\frac{V_{out}}{V_{in}}\right| = \frac{1}{\sqrt{1+(\omega R_3 C_1)^2}} \cdot A_v \qquad (6.3)$$

in which A_v is the gain of the non-inverting amplifier and ω is the frequency, in rad/s.

Considering that the cutoff frequency is the frequency for which the frequency response modulus is equal to $\frac{1}{\sqrt{2}}$, then

$$\omega_c = \frac{1}{R_3 C_1} \quad \text{or} \quad f_c = \frac{1}{2\pi R_3 C_1} \qquad (6.4)$$

6.1.2 High-pass Filter

For the high-pass filter of Figure 6.7, notice that as the frequency increases, the capacitor impedance decreases, so that at high frequencies the output voltage of the circuit equals the input voltage.

It can be shown that the frequency response expression of this circuit is given by

$$\left|\frac{V_{out}}{V_{in}}\right| = \frac{\omega R_3 C_1}{\sqrt{1+(\omega R_3 C_1)^2}} \cdot A_v \qquad (6.5)$$

Note that, in fact, as $\omega \to \infty$, the gain tends to A_v, and that the cutoff frequency is given by the same expression as for LPF

$$\omega_c = \frac{1}{R_3 C_1} \quad \text{ou} \quad f_c = \frac{1}{2\pi R_3 C_1} \qquad (6.6)$$

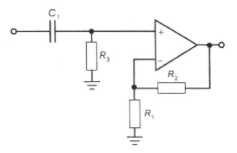

Figure 6.7 First-order HPF circuit.

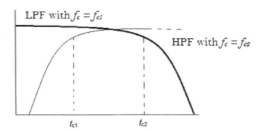

Figure 6.8 Frequency response of the BPF: associating an LPF and an HPF.

6.1.3 Band-pass Filter

The first-order circuits of the LPF and HPF have been shown. Using the two circuits in series, it is possible to construct a bandpass circuit, which allows a frequency response which is the result of the intersection of the two filters (Padilha, 1993), as shown in Figure 6.8.

Thus, the resulting circuit is shown in Figure 6.9.

The expressions for the cutoff frequencies are given by

$$f_{c1} = \frac{1}{2\pi R_1 C_1} \quad \text{and} \quad f_{c2} = \frac{1}{2\pi R_2 C_2}. \tag{6.7}$$

And the central frequency is given by the geometric mean of the two values

$$f_c = \sqrt{f_{c1} \cdot f_{c2}}. \tag{6.8}$$

6.2 Second-order Filters

In this section, we show the generic expressions for the frequency response of 2nd order filters, which involve the factors H_o (maximum gain), Q (quality factor) and ω_c (cutoff frequency).

6.2 Second-order Filters

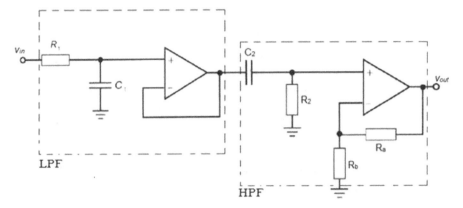

Figure 6.9 First-order BPF circuit.

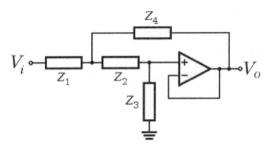

Figure 6.10 Sallen-Key topology.

In addition, the Sallen-Key circuit topology is shown (Figure 6.10), it is widely used in filter construction, due to its flexible and simple implementation (Sedra and Smith, 2004).

It is up to the student to demonstrate that the function of the gain is given by

$$\frac{V_o}{V_i} = \frac{Z_3 Z_4}{Z_1 Z_2 + Z_4(Z_1 + Z_2) + Z_3 Z_4} \quad (6.9)$$

Also, the multiple feedback loop topology of Figure 6.11 is shown.

It is up to the student to demonstrate that the function of the gain is given by

$$\frac{V_o}{V_i} = \frac{Y_1 \cdot Y_3}{Y_3 Y_5 + Y_4(Y_1 + Y_2 + Y_3 + Y_5)} \quad (6.10)$$

in which Y_i is the admittance, given by the inverse of the impedance $\left(Y = \frac{1}{Z}\right)$.

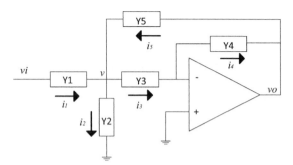

Figure 6.11 Multiple Feedback Loop circuit.

Thus, by using the correct element, it is possible to equate the generic gain function with the transfer function of the second-order filters, a process shown in the following sections.

6.2.1 Low-pass Filter

The second-order LPF has the transfer function given by the following expression (Padilha, 1993)

$$\frac{V_o}{V_i} = H(s) = \frac{H_0 \cdot \omega_0^2}{s^2 + \frac{\omega_0}{Q}s + \omega_0^2} \qquad (6.11)$$

in which:
H_o is the theoretical maximum filter gain;
ω_o is the cutoff frequency;
Q is the filter quality factor;
and s represents the product $j\omega$, in which ω is the frequency in rad/s.

Show that this expression tends to zero for high frequencies, tends to H_o for low frequencies, and has value Q at the cutoff frequency.

For the Sallen-Key topology, in order to get an LPF, one must match the expressions (Sedra and Smith, 1993)

$$\frac{Z_3 Z_4}{Z_1 Z_2 + Z_4(Z_1 + Z_2) + Z_3 Z_4} = \frac{H_0 \cdot \omega_0^2}{s^2 + \frac{\omega_0}{Q}s + \omega_0^2} \qquad (6.12)$$

And the impedances Z_i must be replaced by the impedances of resistors ($Z = R$) and capacitors ($Z = 1/sC$). It can be shown that a possible (but not unique) Sallen-Key configuration of the LPF is as shown in Figure 6.12 (Padilha, 1993).

6.2 Second-order Filters

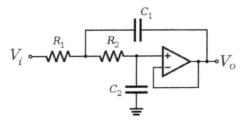

Figure 6.12 Second-order LPF – Sallen-Key.

in which

$$Z_1 = R_1, \quad Z_2 = R_2, \quad Z_3 = \frac{1}{sC_2}, \quad Z_4 = \frac{1}{sC_1}$$

thus, the cutoff frequency, in Hz, is given by

$$f_c = \frac{1}{2\pi\sqrt{R6_1 R_2 C_1 C_2}} \quad (6.13)$$

the quality factor (unitless) is given by

$$Q = \frac{\sqrt{R_1 R_2 C_1 C_2}}{C_2(R_1 + R_2)} \quad (6.14)$$

and

$$\frac{f_c}{Q} = \frac{(R_1 + R_2)}{2\pi C_1 R_1 R_2} \quad (6.15)$$

For the multiple feedback topologies, just as in the Sallen-Key case, it is necessary to match the expressions of the transfer function.

$$\frac{Y_1 \cdot Y_3}{Y_3 Y_5 + Y_4(Y_1 + Y_2 + Y_3 + Y_5)} = \frac{H_0 \cdot \omega_0^2}{s^2 + \frac{\omega_0}{Q}s + \omega_0^2} \quad (6.16)$$

Therefore, choosing $Y_1 = \frac{1}{R_1}$; $Y_3 = \frac{1}{R_3}$; $Y_5 = \frac{1}{R_5}$; $Y_2 = = sC_2$; $Y_4 = sC_4$, one comes to the following expression for the transfer function

$$H(s) = \frac{-\frac{1}{R_3 R_5 C_4 C_2}}{s^2 + s\frac{1}{C_2}\left(\frac{1}{R_1} + \frac{1}{R_3} + \frac{1}{R_5}\right) + \frac{1}{R_3 R_5 C_4 C_2}} \quad (6.17)$$

and the topology seen in Figure 6.13

94   Active Filters

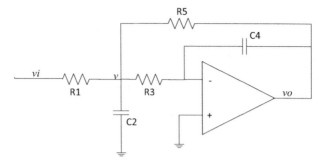

Figure 6.13   Second-order LPF – multiple feedback circuit.

From the previous equation, we observe that

$$\omega_0^2 = \frac{1}{R_3 R_5 C_4 C_2} \rightarrow f_o = \frac{1}{2\pi\sqrt{R_3 R_5 C_4 C_2}} \quad (6.18)$$

$$H_0 = -\frac{R_5}{R_1} \quad (6.19)$$

$$\frac{\omega_0}{Q} = \frac{1}{C_2}\left(\frac{1}{R_1} + \frac{1}{R_3} + \frac{1}{R_5}\right) \quad (6.20)$$

Example: Design an LPF with a cutoff at 1 kHz, $Q = 1/\sqrt{2}$ and unitary maximum gain. Use the multiple feedback topology.

Solution:

Considering $f_0 = 1$ kHz and $Q = \frac{1}{\sqrt{2}}$, and $H_0 = 1$, simplifying the expressions, we obtain:

$$\omega_o = 2000\pi \text{ rad/s},$$

$$\text{For } |H_0| = 1 = \left|-\frac{R_5}{R_1}\right| \therefore R_5 = R_1$$

$$\frac{\omega_0}{Q} = \frac{1}{C_2}\left(\frac{1}{R_1} + \frac{1}{R_3} + \frac{1}{R_5}\right) \therefore \frac{2000\pi}{\frac{1}{\sqrt{2}}} = \frac{1}{C_2}\left(\frac{1}{R_5} + \frac{1}{R_3} + \frac{1}{R_5}\right)$$

$$2000\pi\sqrt{2} = \frac{1}{C_2}\left(\frac{2}{R_5} + \frac{1}{R_3}\right)$$

Selecting $C_2 = 1$ nF and $R_5 = 1$ MΩ, we obtain

$$R_3 = 145 \text{ k}\Omega$$

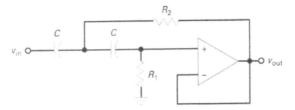

Figure 6.14 Second-order HPF – Sallen-Key.

and, from the expression $\omega_0^2 = \frac{1}{R_3*R_5*C_4*C_2}$, we get

$$C_4 = 175 \text{ pF}.$$

6.2.2 High-pass Filter

The second-order HPF has the transfer function given by the following expression (Padilha, 1993)

$$\frac{V_o}{V_i} = H(s) = \frac{H_0 \cdot s^2}{s^2 + \frac{\omega_0}{Q}s + \omega_0^2} \qquad (6.21)$$

Show that this expression tends to H_o for high frequencies, tends to zero for low frequencies, and is equal to Q at the cutoff frequency.

Again, for the Sallen-Key topology, in order to obtain an HPF, one must match the expressions:

$$\frac{Z_3 Z_4}{Z_1 Z_2 + Z_4(Z_1 + Z_2) + Z_3 Z_4} = \frac{H_0 \cdot s^2}{s^2 + \frac{\omega_0}{Q}s + \omega_0^2} \qquad (6.22)$$

Thus, it can be shown that a possible HPF Sallen-Key configuration is as seen in Figure 6.14.

Thus, it can be shown that the cutoff frequency, in Hz, is given by

$$f_c = \frac{1}{2\pi C \sqrt{R_1 R_2}} \qquad (6.23)$$

and the quality factor is

$$Q = \frac{1}{2}\sqrt{\frac{R_1}{R_2}}. \qquad (6.24)$$

For the multiple feedback topology, it is necessary to match the expressions of the transfer function

$$\frac{Y_1 \cdot Y_3}{Y_3 Y_5 + Y_4(Y_1 + Y_2 + Y_3 + Y_5)} = \frac{H_0 \cdot s^2}{s^2 + \frac{\omega}{Q}s + \omega^2} \qquad (6.25)$$

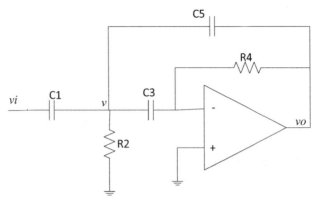

Figure 6.15 Second-order HPF circuit – multiple feedback topology.

Therefore, the following values are chosen: $Y_1 = sC_1$; $Y_3 = sC_3$; $Y_5 = sC_5$; $Y_2 = \frac{1}{R_2}$; $Y_4 = \frac{1}{R_4}$, which yield the following transfer function

$$H(s) = \frac{-s^2 \frac{C_1}{C_5}}{s^2 + s\frac{1}{R_4 C_3 C_5}(C_1 + C_2 + C_3) + \frac{1}{R_2 R_4 C_3 C_5}} \quad (6.26)$$

For which the circuit topology is shown in Figure 6.15.
Following the same procedure as for the LPF, we have

$$\omega_0^2 = \frac{1}{R_2 R_4 C_3 C_5} \quad (6.27)$$

$$H_0 = -\frac{C_1}{C_5} \quad (6.28)$$

$$\frac{\omega_0}{Q} = \frac{1}{R_4 C_3 C_5}(C_1 + C_3 + C_5) \quad (6.29)$$

6.2.3 Band-pass Filter

The second-order FPF has the transfer function given by the following expression (Smith, 2004)

$$\frac{V_o}{V_i} = H(s) = \frac{H_0 \frac{\omega_0}{Q} s}{s^2 + \frac{\omega_0}{Q} s + \omega_0^2} \quad (6.30)$$

Show that this expression tends to H_o for high frequencies, tends to zero for low frequencies, and to Q at the cutoff frequency.

6.2 Second-order Filters

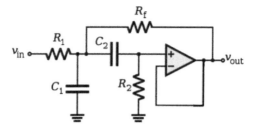

Figure 6.16 Second-order BPF circuit – Sallen-Key.

For the Sallen-Key topology, in order to get a BPF, one must match the expressions:

$$\frac{Z_3 Z_4}{Z_1 Z_2 + Z_4(Z_1 + Z_2) + Z_3 Z_4} = \frac{H_0 \frac{\omega_0}{Q} s}{s^2 + \frac{\omega_0}{Q} s + \omega_0^2} \quad (6.31)$$

Unlike in the previous cases, it will be necessary to add at least one extra element to make a voltage divider of the input signal using a capacitor C_1, as shown in Figure 6.16. Then we obtain one possible Sallen-Key BPF:

Thus, the transfer function is given by

$$H(s) = \frac{\frac{s}{R_1 C_1}}{s^2 + \left(\frac{1}{R_1 C_1} + \frac{1}{R_2 C_1} + \frac{1}{R_2 C_2}\right) s + \frac{R_1 + R_f}{R_1 R_f R_2 C_1 C_2}} \quad (6.32)$$

Thus, it can be shown that the center frequency, in Hz, is given by

$$f_c = \frac{1}{2\pi} \sqrt{\frac{R_f + R_1}{C_1 C_2 R_1 R_2 R_f}} \quad (6.33)$$

The quality factor is given by

$$Q = \frac{\sqrt{\frac{R_f + R_1}{C_1 C_2 R_1 R_2 R_f}}}{\frac{1}{R_1 C_1} + \frac{1}{R_2 C_1} + \frac{1}{R_2 C_2}} \quad (6.34)$$

Notice that for the BPF, the quality factor represents the relation between the center frequency and the -3 dB bandwidth, so the bandwidth is given by (Smith, 2004)

$$BW = \frac{1}{R_1 C_1} + \frac{1}{R_2 C_1} + \frac{1}{R_2 C_2} \quad (6.35)$$

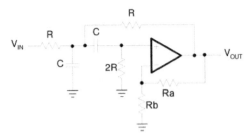

Figure 6.17 Second-order BPF circuit – Sallen-Key with feedback gain.

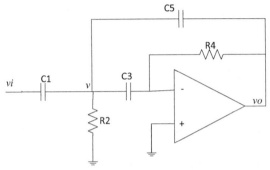

Figure 6.18 Second-order BPF circuit – multiple feedback topology.

Note also that it is possible to produce a variable gain BPF (and quality factor control) by adding the resistors R_a and R_b of Figure 6.17.

In this case, the gain is given by $G = \left(1 + \frac{Ra}{Rb}\right)$, and $Q = \left(\frac{1}{3-G}\right)$.

For the multiple feedback topology, it is necessary to match the expressions of the transfer function:

$$\frac{Y_1 \cdot Y_3}{Y_3 Y_5 + Y_4(Y_1 + Y_2 + Y_3 + Y_5)} = \frac{H_0 \frac{\omega_0}{Q} s}{s^2 + \frac{\omega_0}{Q} s + \omega_0^2} \quad (6.36)$$

Therefore, the following values are chosen: $Y_1 = \frac{1}{R_1}$; $Y_3 = sC_3$; $Y_5 = sC_5$; $Y_2 = sC_2$; $Y_4 = \frac{1}{R_4}$, yielding the following transfer function

$$H(s) = \frac{-s\left(\frac{1}{R_1 C_5}\right)}{s^2 + s\frac{1}{C_5 C_3 R_4}(C_2 + C_3 + C_5) + \frac{1}{R_1 R_4 C_3 C_5}} \quad (6.37)$$

For which the circuit topology is shown in Figure 6.18.

And:

$$\omega_0^2 = \frac{1}{R_1 R_4 C_3 C_5} \quad (6.38)$$

$$H_0 \frac{\omega_0}{Q} = -\frac{1}{R_1 C_5} \quad (6.39)$$

$$\frac{\omega_0}{Q} = \frac{1}{C_5 C_3 R_4}(C_2 + C_3 + C_5) \quad (6.40)$$

Note, nonetheless, that for either topology, another way to design a BPF is to associate an LPF and an HPF in series.

6.2.4 Band-stop Filter

The second-order BSF has the following transfer function (Padilha, 1993)

$$H(s) = \frac{H_0(s^2 + \omega_0^2)}{s^2 + \frac{\omega_0}{Q}s + \omega_0^2} = \frac{H_0 s^2}{s^2 + \frac{\omega_0}{Q}s + \omega_0^2} + \frac{H_0 \omega_0^2}{s^2 + \frac{\omega_0}{Q}s + \omega_0^2} \quad (6.41)$$

It can be seen that this function is the sum of an LPF and an HPF, with the same central frequency, which can be implemented with the use of an adder. And, with some mathematical manipulations, it can be shown that

$$\frac{H_0\left(s^2 + \omega_0^2\right)}{s^2 + \frac{\omega_0}{Q}s + \omega_0^2} = H_0 - \frac{H_0 \frac{\omega_0}{Q} s}{s^2 + \frac{\omega_0}{Q}s + \omega_0^2} \quad (6.42)$$

which can be implemented more easily and steadily using an amplifier with gain H_0 and BPF, as seen in Figures 6.19 and 6.20.

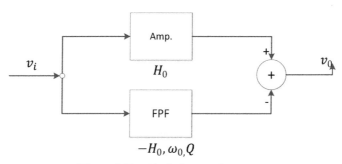

Figure 6.19 Block diagram for a BSF.

100　*Active Filters*

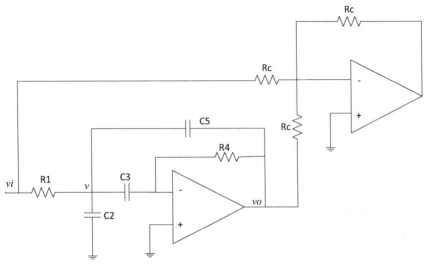

Figure 6.20 Band-stop filter topology.

7
Characterization of Operational Amplifiers

To understand the procedures for measuring the OpAmp characteristics (determined by Lewis Method, or Loop of Tests), readers should be acquainted with the test circuits of Figures 7.1 and 7.2 as well as with the definitions of the parameters. The Op-Amp marked with the acronym DST in both figures is the device subject to test.

The procedures to be described here are accomplished through the circuit of Figure 7.1.

For convenience, the numbers inside circles are referred as terminals, that is, terminal 4, terminal 9 etc. in the following discussion. The relays (switches) are assumed closed in regular operation, except if otherwise stated.

7.1 Extraction of the Offset Voltage (V_{os})

The control voltage, Vc, in terminal 7, should be set to zero. Then the auxiliary amplifier automatically adjusts its output voltage to set the DST output to zero. According to the offset voltage definition, the input voltage V_N assumes V_{os} value, so that the output voltage becomes $1000\, V_{os}$. Therefore, 1.0 V measured at the DST output means that V_{os} is 1.0 mV.

To measure each current separately, the following procedure is accomplished. After extracting V_{os}, the switches K_2 and K_3 are alternately set open and the variations in V_N are calculated. Reading $1000\, V_N$ gives the variation value $\Delta V = (1000) \cdot (I_B) \cdot 10 \text{ k}\Omega$ at the loop output. Thus, if I_B is 100 μA, the measured variation is 1.0 V.

102 *Characterization of Operational Amplifiers*

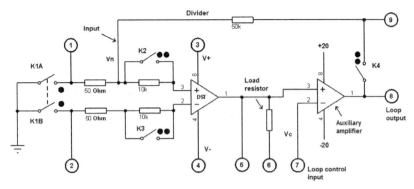

Figure 7.1 Test circuit for the extraction of DST parameters.

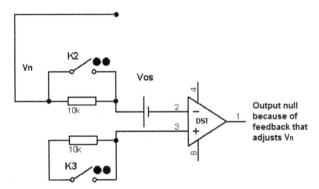

Figure 7.2 Test circuit for the extraction of DST parameters.

7.2 Extraction of Total Bias Current (I_B Total)

At first, the values V_{N2} and V_{N3} are measured one after the other and the difference $V_{N2} - V_{N3}$ is calculated. The variation at the output is given by (1000) I_B *total* · 10 kΩ.

Alternatively, I_B *total* can be extracted by opening the switches K_1 and K_4, connecting terminals 1 and 2 and measuring I_B *total* with an amperimeter. However, since this open loop method is not enough accurate, the previously described procedure has been adopted as an industry standard.

7.3 Extraction of Offset Current (I_{os})

After measuring V_{os}, as described in Section 7.1, the value of V_{os} (in the case the 10 kΩ resistors are not short-circuited by switches K_2 and K_3) renamed

V_{os} (10 kΩ), is also measured and the difference between both values is determined. The multiplying factor is the same used for measuring the other currents: 1000 (I_{os}) (10 kΩ).

7.4 Extraction of the Gain

The test in low gain AC amplifiers, as audio amplifiers, comprises the application of a small signal at the input and the measurement of a large signal at the output (Silva, 2013).

On the contrary, performing the test in high gain DC amplifiers, the inverse procedure applies: the output signal is allowed to sweep through a wide DC range and the resulting DC variation at the input is measured.

For instance, let us suppose it is required to measure the average DC gain through the output range from −10 V to +10 V, which is a regular test for the case of +/− 15 V supply voltages across the DST.

Since the gain is always specified for a load resistor, terminal 6 is grounded and the control voltage is set to $V_c = -10$ V. The auxiliary amplifier forces −10 V at the DST output and V_N is, thus, measured.

In the next step, V_c is changed to +10 V, so that the DST output reaches +10 V. The variation in V_N is measured. If the gain is equal to 100,000 and the total output voltage variation is equal to 20 V (−10 V to +10 V), one should obtain an input voltage variation of (1000). (20/100,000) = 200 mV. It should be noted that a positive variation (increasing voltage) at the output leads to a negative variation (decreasing voltage) at the input. Moreover, the gain is always tested with a load resistor, often of 2 kΩ.

To evaluate the Op-Amp transfer function linearity, the gain test is sometimes performed in two stages.

Instead of applying a single range of output signal variation, from −10 to +10 V, it can be split into two ranges: from 0 to +10 V and from 0 to −10 V. This procedure is able to identify instances for which the gain is very high along a certain portion of the output voltage range although very low elsewhere.

7.5 Extraction of the Common Mode Rejection Ratio (CMRR)

In theory, two signals with equal amplitudes, frequencies and phases applied to the Op-Amp inverter and non-inverter inputs should cancel each other

104 *Characterization of Operational Amplifiers*

so that the Op-Amp output signal is zero (Millman and Halkias, 1972). In practice, however, a small signal appears at the output, which is specified with respect to the maximum magnitude of the gain in terms of attenuation (or rejection). The Op-Amp ability to reject such small signals is the common-mode rejection, which is quantified by the figure of merit named Common Mode Rejection Ratio – CMRR, usually expressed in decibels. In general, CMRR values can range up to 90 dB.

If the input circuit in Figure 7.1 is regarded as a bridge, high values of CMRR can be measured provided the resistors are carefully matched. The method described as follows exempts the use of matched pairs of precision resistors.

To determine the CMRR, a shift of V_{CM} volts is applied above the nominal values of V^-, V^+, and V_c and the variation in V_N is thus measured.

Sometimes the CMRR is extracted by simulating source impedances of 10 kΩ, that is, with the switches K_2 and K_3 set open. In this case, any variation in I_{os} concurs to the total variation in V_N.

7.6 Extraction of the Power Supply Rejection Ratio (PSRR)

For this test, V_c is set equal to zero. The magnitudes of both supply voltages are adjusted to minimum values and V_N is measured. Next, the magnitudes of the supply voltages are raised to their maximum values and the variation in V_N is measured. The PSRR can also be extracted by simulating source impedances of 10 kΩ.

7.7 Extraction of the Output Swing

The output swing is measured at terminal 5. To drive the DST into saturation, a high differential voltage is applied at the input, through one of the following three methods. The more straightforward method consists in opening K_1 and K_4 and applying the voltage between terminals 1 and 2. According to the second method, only K_4 is set open and the voltage is applied to terminal 9. This voltage is attenuated through the ratio 1000:1, as usual, so that 20 V at terminal 9 implies 20 mV at the DST input, which is enough for most operational amplifiers.

The third method is similar to the second, except that all relays (switches) are closed and that 20 V is applied to terminal 9 through the auxiliary amplifier. By setting $V_c = 15$ V, the auxiliary amplifier automatically tries to set 15 V at the DST output, which, however, cannot easily swing up to V^+. Then the auxiliary amplifier saturates around 18 V driving also the DST into saturation. Since the output swing values are specified with respect to a load resistor, terminal 6 must be grounded.

7.8 Extraction of the Short-Circuit Current (I_{sc})

This test comprises the same procedures described in the previous sections, except that instead of using a resistor connected to terminal 6, an ammeter is connected between terminal 5 and ground or between terminal 5 and the opposing supply voltage terminal. As the DST voltage tries to swing according to the input excitation, the DST output is short-circuited by the ammeter and I_{sc} is measured.

7.9 Extraction of the Supply Current

According to usual specifications, the DST output voltage should be set to zero. Therefore, by setting $V_c = 0$, the supply current is measured with the ammeter between V^+ and terminal 3.

7.10 Offset Adjustment

If offset adjustment terminals are available in the amplifier under test, they should be usually connected to V^- by the switches.

V_c voltage should be set to zero and the measurement is performed at the loop output. At least, the measurement should assure that the adjustment range is enough to cancel V_{os}.

It should be noted that the test loop never operates in a simplified form, as explained. Each Op-Amp has its own peculiarities, which require variations over the basic theme. The most common modifications include:

AC Compensation Capacitors – In general, each kind of amplifier presents a different frequency response.

Whenever applying the loop to a particular device, the frequency response curves of both the DST and the auxiliary amplifier can be taken into account in order to define the stabilization circuit to be employed and to preview the necessity of compensation capacitors.

Noise Filter – An RC filter (with a time constant of 1 ms) is usually coupled at the loop output. Often the waveforms observed at the filter output are remarkably more discernible from noise than those obtained directly from the loop output.

Source Resistors – Most general-purpose Op-Amps are tested with source resistors of 10 kΩ. For those with very low input currents, resolution is improved by using resistors of 50 kΩ, 100 kΩ, 1 MΩ or even 10 MΩ.

Loop Test Gain – The most common combination, 50 k/50, gives a gain of 1000. Sometimes the 50-kΩ resistors are split into 45-kΩ and 5-kΩ resistors. For V_{os} extraction, the resistor of 45 kΩ is usually short-circuited by a switch, reducing the gain to 100.

MOSFET switches for K_2 and K_3 implementation – Some devices present extremely low values of I_B and/or V_{os}. If K_2 and K_3 are implemented by REED relays in the test setup, some difficulties may arouse which are related to the relay small-signal features. Thermally generated voltages, leakage currents and vibrations at the relay terminals are typical drawbacks. Although the FET contact resistance is high (100 Ω), its effect is negligible provided the corresponding current is low. Among the advantages of using the MOSFET are: a proper switching, the absence of leakage or offset currents, elimination of multiple-touch problem and of microphony and mitigation of aging.

8

Operational Amplifier Model

The non-ideal Op-Amp differs from the already analyzed ideal Op-Amp in many features (Alley and Atwood, 1973). A very accurate Op-Amp model could be developed by replacing each amplifier transistor and diode by its respective network model. Nevertheless, a typical Op-Amp comprises at least 20 transistors in general.

8.1 Ebers-Moll Complete Model

By choosing the Ebers-Moll complete model (with 13 elements) for the transistors, the resulting circuit presents around 260 elements. It is possible, however, to develop a simpler model – often called macro-model – for the operational amplifier, which is able to accurately represent the real Op-Amp behavior. This behavior is summarized by the following characteristics:

1. Finite input resistance, $R_i \neq \infty$
2. Nonzero output resistance, $R_o \neq 0$
3. Frequency-dependent open-loop gain magnitude – $G(\omega)$. A typical Bode plot for $G(\omega)$ in dB is shown in Figure 8.1. The first noticeable singularity frequency ω_1 is usually known as the dominant pole frequency; the frequency ω_0, in which the $G(\omega)$ curve intercepts the abscissa axis (0 dB axis), is named the unity gain frequency.
4. Frequency-dependent phase shift.
5. Output voltage limitation: $|V_o| < V_{o\,\max}$.
6. Slew rate limit – S_r, defined as the maximum output voltage variation rate attainable by the Op-Amp while connected to the most unfavorable external network. The slew rate[1] is an important parameter for the design

---

[1]Differently from other parameters, the slew rate is not unique, depending upon the feedback configuration in which the Op-Amp takes part. Since the worst case occurs with

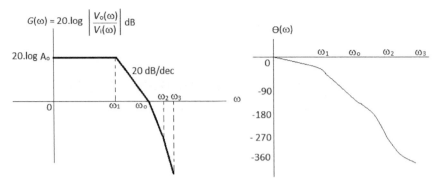

Figure 8.1 Bode plots of the voltage gain magnitude and phase.

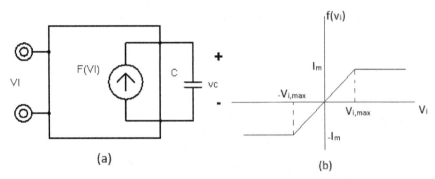

Figure 8.2 Building block for simulating the slew-rate limitation.

of high-speed circuits, as A/D and D/A converters, since it imposes an absolute velocity limit.

The first four characteristics here defined concern the behavior of a linear circuit, being therefore adequate to simulate. The fifth characteristic is typical for voltage limiters and can be simulated with a linear resistor.

In order to simulate the last characteristic, the circuit of Figure 8.2(a) is proposed, in which the voltage-controlled nonlinear current source is represented by the saturation relationship $i_c = f(v_i)$ shown in Figure 8.2(b). Since the output current of the controlled source cannot exceed the value, I_m, then

$$\left|\frac{dv_c(t)}{dt}\right| = \frac{1}{C}|i_c| \leq \frac{I_m}{C} = S_r \tag{8.1}$$

---

the Op-Amp configured as unity gain voltage follower, the slew rate is usually measured in this configuration.

8.1 Ebers-Moll Complete Model

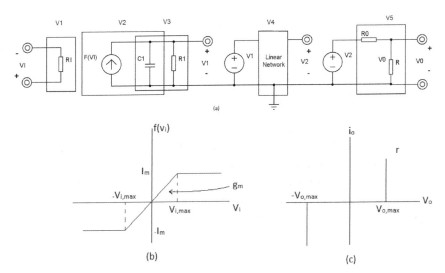

Figure 8.3 Non-ideal Op-Amp macro-model.

From Equation (8.1) it is possible to infer that the slew rate of the circuit in Figure 8.2 is equal to S_r.

According to the previous observations, the circuit of Figure 8.3(a) consists of a possible model for the non-ideal Op-Amp; the nonlinear function in Figure 8.3(b) characterizes the controlled source; the $i - v$ characteristic that represents the nonlinear resistor R is shown in Figure 8.3(c).

It should be observed that this circuit is composed of five stages:

1. Input stage – V_1;
2. Gain and slew rate limitation stage – V_2;
3. Dominant pole stage – V_3;
4. Unity gain stage, for higher-order poles – V_4;
5. Output stage – V_5.

The input stage consists of a resistor R_i, whose resistance is made equal to the input differential resistance specified in the Op-Amp datasheet. The second stage has been reproduced from Figure 8.2 and consists of a nonlinear voltage-controlled current source in parallel with the linear capacitor C_1. Provided the value of R_1 is large enough (to neglect the loading effect in stage V_2) the second stage can be used to simulate any prescribed value of S_r – by choosing I_m and C_1 so that

$$S_r = \frac{I_m}{C_1} \qquad (8.2)$$

110 Operational Amplifier Model

The following analysis shows that one of the available parameters, I_m or C_1, can be chosen so that the macro-model of Figure 8.3 presents the specified DC gain (A_o).

The third stage consists of the parallel association of R_1 and C_1 and is used to simulate the frequency of the dominant pole $\omega_1 = 1/(R_1 C_1)$ of the gain, as it appears in the Bode plot of the magnitude $G(\omega)$, assuming that the magnitude curve decreases with a slope of -20 dB/decade (-6 dB/octave), which is typical for most operational amplifiers. It should be noted that the capacitor C_1 is shared by the stages V_2 and V_3. Again, since there are two parameters to be dimensioned, R_1 and C_1, any convenient value may be assigned to R_1 (provided it is enough large to avoid significantly loading the stage V_2) and the value of C_1 is thus determined by:

$$C_1 = \frac{1}{\omega_1 R_1} \tag{8.3}$$

If the dominant pole frequency ω_1 is not provided by the manufacturer, it can be approximately determined from the unity gain frequency ω_0 (usually informed in the datasheet) using the equation

$$\omega_1 = \frac{\omega_0}{A_0} \tag{8.4}$$

To derive this relationship, it should be observed that if $\omega_2 > \omega_0$, then at $\omega = \omega_0$ most of the current of source $f(v_i)$ flows across C_1, so that

$$V_1 \cong g_m V_i \frac{1}{j\omega_0 C_1} \tag{8.5}$$

in which V_1 and V_i are phasors related to $v_1(t)$ and $v_i(t)$, respectively. Therefore,

$$\left|\frac{V_1}{V_i}\right| = \frac{g_m}{\omega_0 C_1} = 1 \tag{8.6}$$

Now, substituting the value $g_m = A_0/R_1$ (from 8.7, to be derived next) and the value $C_1 = 1/\omega_1 R_1$ (from 8.3) into (8.5) and solving for ω_1, Equation (8.4) is obtained.

The fourth stage is a grounded linear two-port network with unity gain (DC), which is designed to simulate the higher-order poles of the gain, as they appear in the Bode plots of the magnitude $G(\omega)$ and of the phase $\theta(\omega)$. Many RC networks can be applied for this purpose, the associated circuit parameters being determined by classical approximation and optimization techniques. In

8.1 Ebers-Moll Complete Model 111

fact, if V_4 is selected as a minimum phase[2] linear network, from the Hilbert relationship follows that $G(\omega)$ e $\theta(\omega)$ are not independent.

Indeed, given the magnitude function $G(\omega)$, the only associated phase function $\theta(\omega)$ can be calculated and *vice-versa*. This is a significant remark since the transfer functions of most operational amplifiers are of the minimum phase kind; therefore, there is no need of simulating both the gain magnitude and phase characteristics, but only the more important one, the other automatically following from the Hilbert transform relationship.

The last stage consists of a linear and a nonlinear resistor. The value of the linear resistor R_o is equal to the output resistance specified by the Op-Amp manufacturer. The $i - v$ characteristic representing the nonlinear resistor R is like that shown in Figure 8.3(c), in which the limit voltage is equal to the Op-Amp peak voltage $V_{o\,max}$, usually informed in datasheets. It should be noted that the nonlinear resistor behaves like an open circuit for all voltage magnitudes below this limit, that is, for $|v_o| < V_{o\,max}$. Therefore, the last stage is used to simulate both the output resistance and the Op-Amp output voltage limitation.

Now let us determine the parameters I_m and g_m of the nonlinear voltage-controlled current source in Figure 8.3(b).[3] Observing that the DC gain of stage V_4 is equal to unit, then $v_o(\text{DC}) = v_1(\text{DC}) = (g_m v_i)R_1$ in the region of interest; i.e., $-V_{o\,max} \leq v_o \leq V_{o\,max}$. Therefore,

$$g_m = \frac{\frac{v_o(\text{DC})}{v_i(\text{DC})}}{R_1} = \frac{A_0}{R_1} \tag{8.7}$$

in which A_0 is the specified DC gain in an open loop. The value of I_m is obtained by substituting C_1 from 8.3 into 8.2

$$I_m = \frac{S_r}{\omega_1 R_1} \tag{8.8}$$

Finally, dividing (8.8) by (8.7) gives

$$V_{i\,max} = \frac{I_m}{g_m} \tag{8.9}$$

---

[2] A linear two-port network is minimum phase if its transfer function $T(s)$ does not present zeros in the right half plane.

[3] Since the circuit of Figure 8.3(a) is strictly a "black box" model, the parameters I_m and g_m are not related to the Op-Amp transistors. In fact, the choice of these parameters is not unique, but depends upon the choice of the parameter R_1.

112 Operational Amplifier Model

Assuming that the voltage-controlled voltage sources connected to stages V_4 and V_5 are used only for buffering purposes (both sources present unity coefficients), all parameters of the Op-Amp model in Figure 8.3 can be determined from datasheet specifications, as well as from experimental measurements.

To demonstrate the model usefulness, let us consider the following example:

Example: Find a macro-model for the operational amplifier μA 741 with the following specified parameters (typical parameters obtained from the manufacturer datasheet):

Open loop DC gain: $A_0 = 8,35 \times 10^5 = 118$ dB
Differential input resistance: $R_i = 2$ MΩ
Output resistance: $R_o = 75$ Ω
Output peak voltage: $V_{o\,max} = 15$ V
Slew rate: $S_r = 0,5$ V/μs $= 0,5 \times 10^6$ V/s
Dominant pole frequency: $\omega_1 = 2\pi(1,4) = 8,8$ rad/s

Solution: To determine the Op-Amp macro-model parameters one can choose arbitrarily (provided the value is enough large).[4]

$$R_1 = 835 \text{ k}\Omega$$

Therefore:

$$g_m = \frac{A_0}{R_1} = \frac{8.35 \times 10^5}{835 \text{ k}\Omega} = 1 \text{ S}$$

$$I_m = \frac{S_r}{\omega_1 R_1} = \frac{0.5 \times 10^6}{(8.8)(835 \text{ k}\Omega)} = 68 \text{ mA}$$

$$V_{i\,max} = \frac{S_r}{\omega_1 A_0} = \frac{0.5 \times 10^6}{(8.8)(8.35 \times 10^5)} = 68 \text{ mV}$$

Now, conceive the two-port network of unity gain comprised by stage V_4, to simulate the dependence upon frequency of the gain magnitude and phase, for $\omega > \omega_2$. For this example, the four-section ladder network of Figure 8.4(a) has been chosen.

Since this network is of the minimum phase kind, its parameters can be determined in order to comply with either the gain magnitude characteristic

---

[4]To simplify calculations, R_1 has been chosen so that $g_m = 1$.

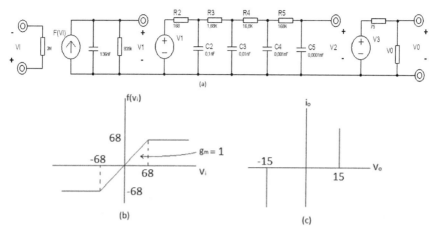

Figure 8.4 µA741 Op-Amp macro-model.

or the phase characteristic of the Op-Amp, which have been specified or measured previously (taking into account that the DC gain is included in stage V_2). In this example, the phase characteristic $\theta(\omega)$ has been chosen and the resulting circuit is shown in Figure 8.4(a).

It should be emphasized that this circuit is not unique. The reader with reasonable experience in filter design and in optimization through computer tools is certainly able to design a circuit so good or even better than this. The fourth-order circuit of Figure 8.4 has been designed to present a fourth-order pole at the frequency $\omega_1 = 2\pi \times 9 \times 10^6$ rad/s, effectively totalizing a phase of $-180°$ at this frequency.

To avoid the use of optimization using computer tools, each capacitance value is a tenth of the left one, producing a negligible loading effect. The transfer function is, thus, approximately equal to the product of the four first-order transfer functions, that is,

$$\frac{V_2(j\omega)}{V_1(j\omega)} = \frac{1}{(1+j\omega R_2 C_2)(1+j\omega R_3 C_3)(1+j\omega R_4 C_4)(1+j\omega R_5 C_5)}$$

It should be noted that for $\omega = 0$, $V_2(0)/V_1(0) = 1$; the DC gain is thus unity. Assuming that $C_2 = 100$ pF, $C_3 = 10$ pF, $C_4 = 1$ pF e $C_5 = 0,1$ pF, the fourth order pole at 9 MHz is introduced by choosing

$$R_2 = \frac{1}{\omega C_2} = \frac{1}{(2\pi)(9 \times 10^6)(10^{-10})} = 168 \, \Omega$$

$$R_3 = \frac{1}{\omega C_3} = \frac{1}{(2\pi)(9 \times 10^6)(10^{-11})} = 1.68 \text{ k}\Omega$$

$$R_4 = \frac{1}{\omega C_4} = \frac{1}{(2\pi)(9 \times 10^6)(10^{-12})} = 16.8 \text{ k}\Omega$$

$$R_5 = \frac{1}{\omega C_5} = \frac{1}{(2\pi)(9 \times 10^6)(10^{-13})} = 168 \text{ k}\Omega$$

8.2 Using the Loop Test

The manufacturers of operational amplifiers usually extract and assure seven main parameters and often extract five secondary parameters of their products. To provide a basis for discussion, the definitions of these parameters follow in the next sections.

8.2.1 The Main Parameters

Input offset voltage (V_{os}) – ideally speaking the output voltage should be equal to zero for null input voltage (DeFrance, 1976). In practice, however, there are small differences between the integrated components inside an Op-Amp. Thus, the offset voltage is the input voltage required to cancel the output voltage, typically in the order of 1 mV or 2 mV. In most cases, V_{os} is due to mismatching between the base-emitter voltage of bipolar transistors, in the amplifier input stage. V_{os} is undesirable in a direct coupling circuit because it is usually amplified by the circuit, producing a large DC error, variable with temperature.

To avoid the effects of the input currents (to be discussed next) V_{os} should be ideally measured with zero source impedance. Here source impedance means the resistance between each input and ground. For test purposes, a low impedance, often equal to 50 Ω, is employed.

Input bias current (I_b) – an ideal Op-Amp does not drain current. In practice, a negligible current flows into each input terminal due to biasing (Millman and Halkias, 1972). In the case of operational amplifiers using bipolar transistors, I_b is the base current of the input transistors, typically of 100 nA. Current I_b does not affect circuits with low source impedance.

In the case of high source impedance, however, the voltage drop given by I_b multiplied by the source impedance appears at the amplifier input – this consists of an error similar to that produced by V_{os}, also dependent upon temperature.

Due to the differential stage design, both Op-Amp bias currents, I_{binv} and I_{bninv}, vary with input voltage, their sum, however, is constant. Since then, the parameter usually extracted is the total bias current $I_{btotal} = I_{binv} + I_{bninv}$. The bias current specified in datasheets is half (1/2) the value of I_{btotal}.

Eventually, both input currents may be measured in separate. If they are measured in an open loop, I_{btotal} is randomly distributed into the two inputs, depending upon V_{os}. The extraction of I_b in closed-loop avoids this random division.

Input offset current (I_{os}) – Since the operational amplifiers present differential inputs, the effect of the input currents can be mitigated, that is, partially canceled, provided the currents are equal (Padilha, 1993).

In practice, the input currents cannot be made equal, thus the maximum allowed difference between these currents is specified. The offset current is the difference between the currents flowing into the input terminals, for the condition of zero output. In the case of single input stage operational amplifiers (μA709, μ749), I_{os} depends upon the matching between the gains of the transistors (*beta*). In more complex amplifiers (μ741) I_{os} also depends upon the matching between the current sources feeding the input transistors.

Offset voltage for high impedance (V_{os} **10 kΩ**) – Errors in the DC gain A_0 are completely specified with the set of parameters V_{os}, I_{binv}, I_{bninv} or V_{os}, I_{os}, and I_{btotal}. Using any of these sets, the input common-mode and differential voltages can be determined for any pair of input resistances, either equal or not.

For instance, let us assume that only the input differential voltage for identical input resistances does matter, as in actual applications. For low impedances, V_{os} dominates; for high impedances, the effects of I_{os} dominate.

For intermediate resistances, V_{os} and I_{os} produce effects of the same order. These effects may add or cancel each other, depending on their polarities, statistically uncorrelated.

If there is a sum, the composed effect in the errors is more pronounced than V_{os} and may even overcome the datasheet limit for the offset voltage. Therefore, it is convenient to measure this offset voltage using a defined value for the input (source) impedance, usually, 10 kΩ, to avoid the previously mentioned possibility.

The datasheets state that the value of V_{os} is assured for source resistances below 10 kΩ and that "V_{os} can be defined for the case equal source resistances are introduced in series with the input terminals". It is implied that V_{os} (10 kΩ) is not an independent parameter. It is calculated from the interaction

116 *Operational Amplifier Model*

between the independent parameters V_{os} and I_{os} with resistors connected to the inputs.

Voltage Gain (A_V) – The Op-Amp gain is defined in the same way as any amplifier gain, that is, as the ratio between the output voltage variation and the corresponding input voltage variation. The manufacturer specifies a typical gain, in general, also assuring a minimum value for the gain (Silva, 2013).

Common Mode Rejection Ratio (CMRR) – In the ideal case, an Op-Amp ignores the common-mode signals. In practice, there is a small response to common-mode voltage variations. The ratio between the differential voltage effect and the common-mode voltage effect is known as CMRR (Common Mode Rejection Ratio), usually measured in dB. Typically, Op-Amps present CMRR in the order of 80 up to 100 dB.

Power Supply Rejection Ratio (PSRR) – This relationship evaluates the Op-Amp ability to ignore the supply voltage variations. To extract this parameter, the variation in the value V_{os} must be measured while supply voltage is simultaneously modified. The PSRR is defined as the ratio between V_{os} variation and supply voltage total variation. For example, if the supply voltage varies from $+/-20$ V to $+/-5$V, the total variation is $40 - 10 = 30$ V. The PSRR is usually specified in uV/V or sometimes in dB. A typical value of the PSRR is 30 uV/V (90 dB).

8.2.2 The Secondary Parameters

Output Swing – Ideally, the Op-Amp output voltage might swing unlimitedly despite the supply voltages. Nevertheless, the magnitude of the output voltage of actual Operational Amplifiers saturates at 1 volt or 2 volts below the supply voltage magnitude. The saturation level depends upon how many base-emitter junctions and/or saturated transistors are enclosed. The Op-Amp output stage is normally composed of emitter followers in complementary symmetry to achieve a low output impedance. To assure that both NPN and PNP transistors operate properly, the positive and negative swings should be tested using an external resistor as output load.

Output Short-Circuit Current (I_{sc}) – Most Op-Amps comprise a built-in protection circuit for the current. If the output is short-circuited or overloaded, the output current is limited in a safe value, typically 25 mA. Although designed for the same clamping current magnitude, the limiting circuits for both current directions (source or sink mode) are independent and should be tested separately.

Particularly, the current value should not produce an exceedingly fast output voltage variation.

If the current source provides a current value I through capacitor C, the output voltage increases up to a limiting value at the ratio named Slew Rate, as follows:

$$\frac{dV}{dt} = \frac{I}{C}$$

This parameter is expressed in volts per second, varying from 1 V/μs up to 1000 V/μs. The slew rate limit is reached when the current is not enough to charge the capacitor, that is when a large output voltage signal varies too fast.

Another way to understand the Slew Rate is through its relationship with the bandwidth. Overexciting the Op-Amp with a high-frequency sinusoidal large-signal, a triangular waveform appears at its output. The slope of this waveform is the Slew Rate.

8.3 Basic Test Loop for Operational Amplifiers

Fortunately, for test engineers, all Op-Amps are so similar to each other that a single test circuit is able to accomplish all standard tests.

This circuit, shown in Figure 7.1 is the "Basic Test Loop for Op-Amp". The numbered circles in the test loop diagram represent the programmable terminals or Kelvin groups in an automatic tester and the switches are programmable Reed relays. In a simpler manner, the circles may be thought of as terminals and the switches as buttons.

In general, five supply sources are needed for all tests: the sources V+ e V− for the device under test (DST), a loop control voltage and a few sources to bias the auxiliary amplifier, usually providing +/−15 V or +/−20 V.

Assuming that all relays are shut, the loop operation proceeds as follows:

− The auxiliary amplifier inverting input is the loop control terminal.

In general, I_{sc} is tested under the worst conditions. For instance, the input voltage is adjusted in order to produce positive saturation. Next, the output terminal and the negative supply voltage terminal are short-circuited and the resulting current is measured.

Supply Current I_s, I_{sop}, $I+$ **or** I_{cc} − The amplifier supply current is measured with zero output voltage. Since recent Op-Amps do not present ground terminal, the supply current flowing through the V+ terminal is equal to that flowing through the V− terminal. This current can be measured at any

118   *Operational Amplifier Model*

terminal. For ancient Op-Amps presenting the ground terminal, as the uA702, $I+$ and $I-$ must be measured separately.

Power consumption – The power consumption is determined by multiplying the supply current by the total supply voltage. The consumption is assured by the $I+$ test.

Offset adjustment V_{os} (adj) – Some Op-Amps present a pair of terminals for offset adjustment. By connecting and conveniently adjusting a variable resistor between these terminals, it is possible to cancel V_{os}. The influence of each terminal voltage V_{os} (adj) on the value V_{os} is assessed by connecting this terminal with $V-$ terminal and measuring V_{os}.

8.3.1 AC Parameters

Since manufacturers do not usually test the AC parameters, datasheets present only their typical values. However, engineers should understand the meaning of the three most common AC parameters:

Rising time and overshoot – the step response (considering a small signal at the input) consists of a simple test that indicates the passband as well as the amplifier stability under specified conditions. The rising time is related to the passband and the overshoot is a measure of the amplifier stability.

Slew rate – The slew rate, which represents a large signal operation inherent phenomenon, is due to the introduction of capacitors for adjusting the small-signal frequency response. To reduce the required capacitance, these capacitors are usually connected to high impedance nodes of the circuit. Such nodes are biased by current sources. If a 10-volt step should be reproduced, the amplifier deviates from its normal behavior because of the high signal magnitude.

Since the following conditions are met:

- the DST output is connected to the auxiliary amplifier non-inverting input;
- the auxiliary amplifier controls the DST input through the feedback voltage divider;
- the auxiliary amplifier operates with negative feedback due to the inversion produced by the DST in the loop;
- with negative feedback, the auxiliary amplifier continuously adjusts its output voltage in order to keep its input voltages equal;

then the basic rule is stated: the auxiliary amplifier continuously adjusts the loop output so that the DST output follows the control terminal.

The input node voltage (V_n) is always very close to 1/1000 (a thousandth) of the loop output voltage. In other words, the circuit closed-loop gain is 1000, because any value V_n is multiplied by 1000 at the output.

This gain is very useful for measuring the small voltages involved in the process. The tests described as follows demonstrate the versatility and simplicity of the basic test loop. Nevertheless, before discussing how to measure each parameter through the test loop, the meaning of V_n for each configuration of K_2 and K_3 relays should be evaluated.

8.3.2 V_n Equations

The input circuit of the DST does not show the 50-ohm resistors since their effects are negligible in this analysis.

V_{os} is modeled as an external small voltage source. The bias currents flow through both inputs. Adjusting V_n by the feedback so that the DST output voltage is equal to zero and assuming that V_{os} is an external voltage, there is no voltage difference between the Op-Amp inputs. Such a circuit makes it possible to write equations for the four possible configurations of K_2 and K_3 relays.

1. K_2 and K_3 are closed
   This is the simpler case: $V_{n1} = V_{os}$
   The biasing currents do not significantly affect the value V_n since there is no source resistance.
2. K_2 is closed and K_3 is open
   $V_{n2} = V_{os} - I_B(ninv).(10\ \text{k}\Omega)$.
   With only K_3 open, V_n is a composite voltage, comprising V_{os} and $I_B(ninv)$, which suggests a way of testing $I_B(ninv)$: measuring V_{n2} and subtracting V_{n1} from it to obtain $I_B(ninv)$.
3. K_2 is open and K_3 is closed
   $V_{n3} = V_{os} - I_B(inv).(10\ \text{k}\Omega)$.
   Similarly, $I_B(inv)$ is obtained from the difference between V_{n3} and V_{os}.
4. K_2 and K_3 are open
   $V_{n4} = -I_B(ninv).(10\ \text{k}\Omega) + V_{os} + I_B(inv).(10\ \text{k}\Omega)$
   $V_{n4} = V_{os} + [I_B(inv) - I_B(ninv)].(10\ \text{k}\Omega)$
   $V_{n4} = V_{os} + I_{os}(10\ \text{k}\Omega) = V_{os}.(10\ \text{k}\Omega)$
   V_{n4} is the composite offset voltage for high source impedances.

120 Operational Amplifier Model

Each of these four relay configurations consists of an easy way for measuring an important Op-Amp input parameter. Even the measurement of $I_B(total)$, instead of the individual input currents, proves to be easy: V_{n3} is measured and then subtracted from V_{n2}. From the previous equations follows:

$$\Delta V_n = V_{os} - I_B(inv).(10 \text{ k}\Omega) - I_B(ninv).(10 \text{ k}\Omega) - V_{os}$$
$$\Delta V_n = [I_B(inv) - I_B(ninv)].(10 \text{ k}\Omega)$$
$$\Delta V_n = I_B(total).(10 \text{ k}\Omega)$$

9

Oscillators

Oscillators are electronic circuits that are intended to generate, from DC power, an AC power signal. That is, they produce periodic signals at their output, with specific frequencies obtained according to the oscillator configuration. Depending on the type of oscillator, the waveform at its output may be pure or non-sinusoidal, such as triangular, square, pulse, sawtooth, among others.

These electronic systems are important in communications systems because they are used in a variety of applications, such as radio transceivers and signal generators for information transmission.

This chapter introduces the main concepts of the oscillator theory. Initially, the chapter describes the ideal form of oscillation, fundamentals of sinusoidal oscillators, Barkhausen criterion, limiter circuits and the main types of LC oscillators. Finally, the mixer circuit and voltage control oscillator is described.

9.1 Types of Oscillators

In general, oscillators can be classified into two groups: linear or tuned oscillators and non-linear or relaxation oscillators. Linear oscillators are responsible for generating pure sine waves and are designed using transistors and/or amplifiers. Non-linear oscillators generate triangular waves, square, pulse, sawtooth, among others, and are implemented from bistable devices such as logic gates and flip-flops.

It is important to note that, regardless of the type of oscillator, i.e. whether it is linear or non-linear, the output of the oscillator is an alternating voltage due to the correct polarization of its components (amplifier and/or transistor) when it is fed by a direct current.

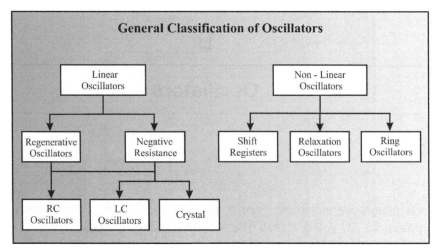

Figure 9.1 Block diagram with overall oscillator classification.

Figure 9.1 the classification of the oscillators is shown, in which it is possible to observe that the linear oscillators are subdivided into regenerative and of negative resistance, whereas the non-linear oscillators can be of the shift register, relaxation or ring structure. The study of these electronic circuits requires the observation of its operating principle, waveform, and frequency of oscillation.

Linear or harmonic oscillators are characterized by producing a sine wave at the output. They are implemented from an active element, usually an amplifier, and a feedback system that feeds a particular output frequency range back to the input. Among this type of oscillator, stand out RC oscillators, LC oscillators, and crystal.

The RC oscillators are so named because they use resistive and capacitive components in their feedback system. These circuits cause a phase shift and provide positive feedback. They are used in applications that require signals with low frequencies and their oscillation frequency is determined by the values of the resistive and capacitive components used in their design. In this type of oscillator, two well-known architectures in the literature are the Wien bridge oscillator and the phase shift oscillator.

The LC oscillators are thus named for using capacitive and inductive components in their feedback system. In this type of oscillator, there is generally a parallel circuit which is periodically fed by voltage pulses in order to maintain the current flowing through it. In its operation, the circuit current

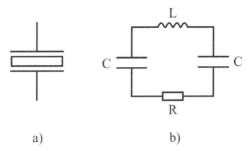

Figure 9.2 (a) Symbol representing the crystal of a piezoelectric material. (b) Equivalent electrical circuit.

travels opposite directions causing the inductors and capacitors to change their energies (Forouhar Farzaneh, 2018).

The LC oscillators, when compared to other types of oscillators, present in their output a waveform that usually have many harmonics and have a narrow tuning range. However, these oscillators have good instability, low immunity to jitter and low phase noise. These oscillators are used in applications that require signals with a higher oscillation frequency when compared to signals produced by RC circuits. The most widespread architectures in the LC oscillator literature are Armstrong, Colpits, and Hartley.

The last type of harmonic oscillator is crystal. The crystal oscillator was created in the 1920s and has the characteristic of using the resonance of vibrating crystal of a piezoelectric material, that is, a material that generates an electrical voltage when subjected to mechanical pressure. To achieve the maximum frequency stability, the crystal oscillator uses quartz crystals.

Figure 9.2(a) illustrates the symbol representing the crystal, while Figure 9.2(b) illustrates its equivalent circuit. The crystal oscillator is formed by the connection of two terminals connected to an internal piezoelectric crystal. The oscillation frequency is defined from the crystal contraction and relaxation times. When subjected to the application of an electric voltage, the crystal contracts and, after a certain time from the crystal construction, release the voltage and, then relaxes. This cycle repeats at a frequency which is the operating frequency of the oscillator.

9.2 The Ideal Oscillator

The ideal oscillator, also known as a resonant tank, is formed by an LC circuit, as shown in Figure 9.3.

124 Oscillators

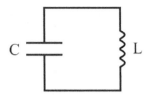

Figure 9.3 Resonant tank circuit.

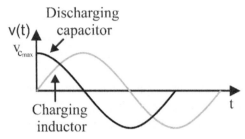

Figure 9.4 Capacitor and inductor charge and discharge behavior in the resonant tank.

Since the oscillator is ideal, the ideal oscillator electric model of Figure 9.3 is lossless. In this case, the electric energy of the capacitor and the magnetic energy of the inductor is conserved when the circuit is excited by a voltage source.

Thus, in order to obtain an ideal oscillation, initially, the capacitor must be charged by means of a source of temporary voltage until reaching its maximum voltage, $v_{C_{max}}$. When the voltage source is disconnected, the capacitor will discharge through the inductor, which consequently begins to charge.

The inductor, which withstands alternating current flow, carries up to a maximum point, $V_{i_{max}}$, which is the point at which the capacitor has zero charge, and after the current passes, the inductor discharges in the reverse direction. Figure 9.4 shows the charge and discharge process of the reactive components and the formation of the ideal sine wave, in which it is possible to observe the 90° phase difference between the current of the inductor and capacitor.

In order to analyze the electrical circuit of the ideal oscillator, Kirchhoff's Voltage Law (KVL) can be applied, which states that the sum of the voltages along the circuit is zero. Therefore, by traversing the circuit clockwise, thus (Riedel, 2008)

$$v_L + v_C = 0, \qquad (9.1)$$

9.2 The Ideal Oscillator

in which V_L and V_C are, respectively, the inductor and capacitor voltage. From Expression 9.1, it can be concluded that

$$L\frac{di}{dt} + \frac{1}{C}\int idt = 0, \qquad (9.2)$$

in which L and C are, respectively, capacitance and inductance.

Deriving and then dividing both of the terms of Expression 9.2 by L,

$$\frac{d^2i}{dt} + \frac{1}{LC}i = 0. \qquad (9.3)$$

The solution of Expression 9.3 is obtained from the roots of its characteristic equation, that is

$$s^2 + \omega_o = 0, \qquad (9.4)$$

in which ω_o is the natural frequency of oscillation of circuit given by

$$\omega_o^2 = \frac{1}{LC}. \qquad (9.5)$$

The solution for Expression 9.4 is

$$i(t) = I_1 e^{-j\omega_o t} + I_2 e^{j\omega_o t}. \qquad (9.6)$$

The roots of the characteristic equation are

$$s_1 = j\omega_o, s_2 = -j\omega_o. \qquad (9.7)$$

In the ideal context, the charge-change cycle stored in the reactive devices would remain constant. However, the actual circuits undergo Joule dissipation, causing the oscillations to be attenuated until all accumulated energy is consumed. The effect of the damped oscillations is illustrated in Figure 9.5. Depending on the oscillation frequency, there may be several oscillation cycles until the energy reaches zero.

In the case of a lossy circuit, the characteristic equation must include a linear term and can be expressed by

$$s^2 + 2\alpha s \omega_o = 0. \qquad (9.8)$$

In this case, considering $\omega_o > \alpha$ the roots of the characteristics equations are given by

$$s_1 = -\alpha + j\sqrt{\omega_o^2 + \alpha^2} \approx -\alpha + j\omega_o. \qquad (9.9)$$

126 Oscillators

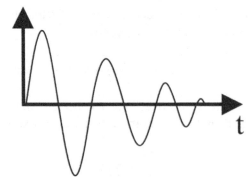

Figure 9.5 Energy dissipation in a feedback loopless oscillator.

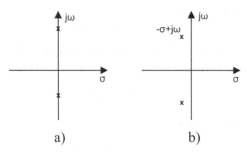

Figure 9.6 Root arrangement of the characteristic equation: (a) ideal oscillator. (b) lossy oscillator.

and

$$s_2 = -\alpha - j\sqrt{\omega_o^2 + \alpha^2} \approx -\alpha - j\omega_o. \tag{9.10}$$

In Figure 9.6 the arrangement of the natural oscillation frequencies in the complex frequency plane is shown, considering the lossless and lossy resonator.

In order to work around the damping of the oscillations caused by the losses in the circuit, it is necessary to include in the oscillator circuit an active element and a feedback of the output signal at the input, in order to be able to restart the cycle and maintain the oscillator output signal.

9.3 Fundamentals of Sinusoidal Oscillators

In the previous section the resonator circuit, which consists of the ideal harmonic oscillator, was approached. In this configuration, it is considered

9.3 Fundamentals of Sinusoidal Oscillators

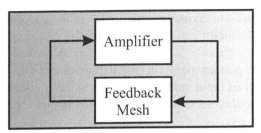

Figure 9.7 Basic structure of a real oscillator.

that there are no losses in the electrical circuit and the output of the oscillator is a pure sine wave.

However, in real situations, it is necessary to consider the intrinsic losses to the electrical circuits. In the case of the resonator, the oscillations are dampened due to the Joule effect present in the equivalent resistors in series with the capacitor and the inductor, which causes the temperature increase and the release of heat in the passage of electric current through the resistor.

Thus, by applying a voltage source to the capacitor of the resonator, the oscillations will begin to occur. However, because of the losses, the output signal of the oscillator loses energy, causing the oscillations to attenuate until they are zero. In this process, depending on the frequency, there can be several oscillations until the consumption of all the energy in the circuit.

In order for the oscillator to have permanent oscillations at its output, that is, oscillations that are constantly repeated, it is necessary to include in its circuit an active element and a feedback system. Figure 9.7 illustrates the basic structure of an oscillator, considering the amplifier as an active element and a generic feedback system (Smith, 2004).

In Figure 9.7, the amplification system, which can be implemented by means of a transistor, for example, must have a high gain in order to compensate for the losses and amplify the return signal, to have at the output of the oscillator, a wave with permanent oscillation.

The feedback is the basis of oscillation. It allows the energy to circulate continuously in the circuit, compensating for its losses. The feedback signal generated by the circuit is proportional to the input signal and, if they are in phase, and if the magnitude of the feedback signal is large enough, a regenerative situation occurs and the oscillator becomes an unstable circuit.

Instability in oscillators is an important factor and is related to mesh gain. In order for the oscillator to work correctly, a balance must be struck between a situation of instability and stability. For this, it is necessary that the mesh

128 Oscillators

gain equals unity. Otherwise, the amplitude of the generated waveform at the output of the oscillator can grow exponentially or decay exponentially until zero.

Another important point in oscillating circuits is the need for non-linearity. When an input voltage is applied to the oscillator circuit, there will be transient oscillations that are stabilized by the use of a non-linear control loop.

When considered a real oscillator, its generated output has harmonic distortions, which consist of sinusoids with multiple frequencies of the fundamental, and is a parameter used to determine the efficiency of the oscillator. This parameter is called Total Harmonic Distortion (THD), usually expressed as a percentage, and is given by (Campos, 2015)

$$THD = \sqrt{\frac{a_1^2 + a_2^2 + a_1^3 + \cdots + a_n^2}{a_0^2}} x100\%. \qquad (9.11)$$

in which $a_1^2, a_2^2, a_1^3, \ldots a_n^2$ are the amplitudes of the nth harmonics and a_0^2 is the amplitude of the fundamental frequency.

9.3.1 Barkhausen Criterion

The Barkhausen criterion, so named for being developed in 1921 by the German physicist Heinrich Georg Barkhausen (1881–1956), describes the stability condition of the oscillators.

Initially, consider Figure 9.8, which consists of an adaptation of Figure 9.7, including the step of the sum and an input signal, x_i. This change is only to make clear the proposal of the Barkhausen criterion. In practice, the oscillators are represented by the diagram in Figure 9.7 and have no input signal (Smith J. R., 1997).

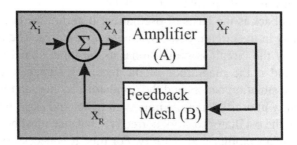

Figure 9.8 Detailed structure of a real oscillator.

9.3 Fundamentals of Sinusoidal Oscillators

The stability condition of the oscillators proposed by Barkhausen is related to the mesh gain. Thus, through Figure 9.8, it is

$$x_f = Ax_A. \tag{9.12}$$

and,

$$x_R = Bx_f = ABx_A. \tag{9.13}$$

The input signal on the amplifier is related to the input signal and the feedback signal by

$$x_A = x_R + x_i. \tag{9.14}$$

Thus,

$$x_A = x_i + ABx_A \Rightarrow x_A[1 - AB] = x_i. \tag{9.15}$$

Replacing Formula 9.13 in Formula 9.15, find the feedback gain, L_R, given by

$$L_R = \frac{x_f}{x_i} = \frac{A}{1 - AB}. \tag{9.16}$$

In this way, the gain of the closed mesh, $M(s)$, is given by

$$M(s) = A(s)B(s). \tag{9.17}$$

In the oscillators, the closed-loop gain is an important parameter because it is through it that the condition of stability, instability or equilibrium between stability and instability is observed.

The closed-loop gain may assume greater value, equal to or less than unity. If it is larger than the unit, there is an instability situation, and in that case, the amplitude of the output oscillations grows exponentially. If the closed-loop gain is less than unity, there is a stability situation, in which the wave amplitude will drop exponentially to zero. An equilibrium situation is reached when the closed-loop gain equals unity, indicating that the amplitude of the oscillator output wave remains constant.

Note that from Formula 9.16, if the loop gain is unitary, the feedback gain is infinite, and the circuit will have finite output for a null input signal, thus characterizing a typical circuit of an oscillator.

The criterion of Barkhausen says that the only circuits present in their output oscillations in permanent regime in a certain frequency f_0 if they meet two conditions. They are:

1. The magnitude of the closed-loop gain should be equal to unity, that is,

$$M(j\omega_o) = A(j\omega_o)B(j\omega_o) = 1. \tag{9.18}$$

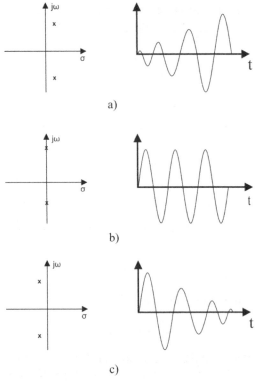

Figure 9.9 Relationship between the poles of the characteristic function and the oscillation behavior: (a) instability and gain greater than unity. (b) balance between instability and stability. (c) stability and gain less than a unit.

2. No phase change in feedback (phase shift 0°), that is,

$$\angle M(j\omega_o) = 0° \qquad (9.19)$$

The essence of Barkhausen's criterion is, from these two conditions, to make the circuit stable by positioning the poles of its characteristic equation exactly on the imaginary axis of the plane of frequencies. Figure 9.9 illustrates the relationship of the oscillation signal with the position of its poles in the plane.

Figure 9.9 shows the three possible situations, that is, stability, instability, and equilibrium between them. In Figure 9.9(a) the instability situation is shown when the mesh gain is greater than one and the poles of the characteristic equation are displaced to the right, which results in the exponential growth of the wave amplitude. In Figure 9.9(c), the output of the oscillator is shown when the circuit is stable, that is, the closed gain is less than unity, resulting in the left detachment of the poles from the characteristic equation and the exponential decay of the amplitude of the oscillation. Finally, in Figure 9.9(b) there is a balance between stability and instability, in which the poles of the characteristic equation are positioned exactly on the imaginary axis, resulting in oscillations with constant amplitude at a given frequency. It is important to note that the output signal of the oscillator will only show constant amplitudes after generating initial oscillations, called starting oscillations.

The starting oscillations are caused by making the mesh gain larger than the unit. In this case, the oscillator is in a state of instability. Next, a nonlinear circuit must be used to control the exponential growth of the amplitude of the starting oscillations. From the nonlinear circuit or limiters, the desired amplitude for the oscillations is defined, as explained below.

9.4 Limiter Circuits

Limiting circuits, also known as limiter circuits, are widely used in electronics. They are intended to limit the voltage levels of an excitation source and are formed by diodes and resistors, in addition to the excitation source.

The behavior of the limiting circuits are observed in their transfer characteristic curve. Figure 9.10 illustrates the transfer curve for an ideal limited circuit. It is observed that the limiter circuit reaps the input voltage according to the following relation:

$$\frac{-A}{K} \leq v_i \leq \frac{+A}{K}. \qquad (9.20)$$

in which $-A$ and $+A$ are, respectively, the lower and upper outlet saturation limits and K is a constant of proportionality.

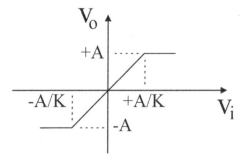

Figure 9.10 Transfer characteristic curve of an ideal limiter.

132 *Oscillators*

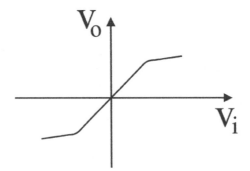

Figure 9.11 Transfer characteristic curve of an real limiter.

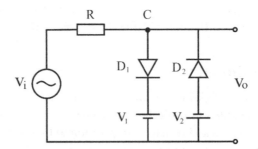

Figure 9.12 Example of a limiter circuit.

In real limiting circuits, the transfer characteristic curve shows a smooth transition between the linear region and the saturation region, as shown in Figure 9.11.

Limiters can be configured to suppress only the upper limit or only the lower limit. If one of these limits is equal to zero, the limiter circuit operates as a half-wave rectifier circuit.

Figure 9.12 shows an example of a limiter circuit, formed by an excitation source (input signal) v_i, one resistor R, two diodes D_1 and D_2 and two sources, v_1 and v_2.

The analysis of the circuit operation can be divided into two stages. Consider initially that the excitation voltage has a positive value. In this case, if the voltage v_1 is greater than the voltage at point C, the diode D_1 will start conducting and the diode D_2 will be in cut. Since the conducting diode is represented by a short circuit and the cutoff diode is an open circuit, the circuit of Figure 9.12 is then represented by the circuit of Figure 9.13.

The output voltage, v_o, of the limiter is initially obtained by input voltage and diode voltage, $v_D = 0,7$. In principle, the input voltage has a value

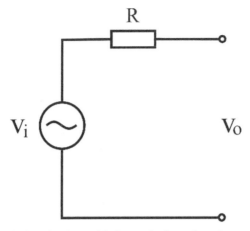

Figure 9.13 Limiter circuit considering excitation voltage has a positive value.

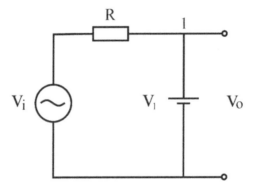

Figure 9.14 Limiter circuit considering excitation voltage has a negative value.

lower than the voltage in 1, that is, $v_1 + v_D$ (Gomes, 1985) (Campos, 2015). Thus, the output voltage follows the shape of the excitation voltage. When the excitation voltage reaches the voltage value of branch 1, the output voltage assumes the value of $v_1 + v_D$ until the input voltage returns to a value lower than the branch voltage and, once again, the output voltage accompanies the input voltage.

The second part of the analysis is done considering the negative half-cycle of the input voltage. In such a case, the diode D_2 will be in conducting while the diode D_1 will be in cut. In this situation, the circuit is reduced to that shown in Figure 9.14. Faced with this configuration, the voltage at point C must be greater than the voltage v_2. The output voltage accompanies the

134 Oscillators

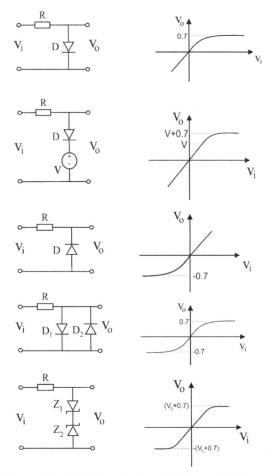

Figure 9.15 Examples of limiting circuits and their transfer characteristic curves.

negative half-cycle of the excitation voltage, but it assumes a constant value and equals the voltage in branch 2, $-(v_1 + v_D)$, until the excitation voltage returns to a value greater than the voltage of branch 2.

Other examples of limiting circuits are shown in Figure 9.15. Following the same analysis above, it is possible to obtain their respective transfer characteristic curves.

The limiter analysis performed here serves as a basis for understanding the limiting circuits used in the oscillators. The following is an example of a limiting circuit used in oscillators.

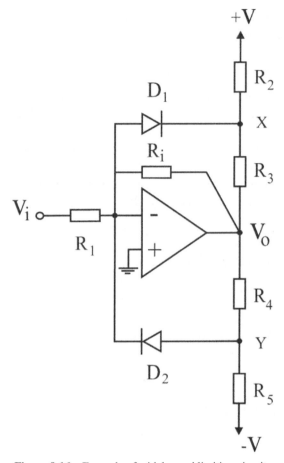

Figure 9.16 Example of widely used limiting circuit.

9.4.1 An Example of a Limiter Circuit Used in Oscillators

As mentioned, oscillator circuits need to meet Barkhausen's criteria to ensure stability. However, furthermore, these circuits make use of limiting circuits. An example of a nonlinear mesh commonly used in oscillators is shown in Figure 9.16 (Smith A. S., 2004).

The non-linear loop circuit consists of an amplifier and a limiter circuit, with two diodes D_1 and D_2, and six resistors R_1, R_2, R_3, R_4, R_5 and R_i.

The idea of the nonlinear mesh is to find the upper and lower thresholds of the output wave amplitude of the oscillator. In the end, it is expected to obtain

136 Oscillators

the transfer characteristic graph of the limiter. To understand the operation of the circuit, consider the following three cases:

Case 1: diodes without conduction

The first step in understanding the operation of the circuit and obtaining the graph that illustrates the transfer characteristic of the limiter is to analyze it without the presence of the diodes D_1 and D_2.

For this, suppose that the input signal, v_i, and the output signal, v_o, are small enough that they are not enough to activate the conduction of the diodes. All input current will pass through resistance R_i. Thus, the relation between the input and output signals is given by:

$$v_o = -G_t v_i, \tag{9.21}$$

in which the limiter gain, G_t, is given by:

$$G_t = -\left(\frac{R_i}{R_1}\right). \tag{9.22}$$

Formula 9.21 suggests a linear relationship between the input voltage and the output voltage. Thus, for this first case, the transfer characteristic is shown in Figure 9.17.

Case 2: Inferior Limit

To determine the lower limit of the oscillator output wave, consider v_X, obtained by the superposition of voltage at point X. The voltage v_X is given

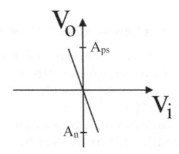

Figure 9.17 Linear relationship between the input voltage and the output voltage of the limiting circuit.

by:
$$v_X = \frac{R_3}{R_2+R_3}V + \frac{R_2}{R_2+R_3}v_o. \quad (9.23)$$

According to Formula 9.21, it is possible to observe that when v_i assumes positive values, v_o assumes negative values, with a difference of amplitude given by the gain of the limiter. Thus, for the positive half-cycle of the input wave, in which it begins with a very small voltage, practically null, it is observed that as the voltage increases, the output voltage, v_o, decreases and, consequently, the voltage v_X becomes increasingly negative.

This process happens until v_X assumes the value necessary for the diode D_1 to conduct. That is,
$$v_X = -v_D, \quad (9.24)$$
in which v_D is the voltage in the diode. In this case,
$$v_o \frac{R_2}{R_2+R_3} = v_X - V\frac{R_3}{R_2+R_3}. \quad (9.25)$$

Consequently,
$$v_o = v_A \frac{(R_2+R_3)}{R_2} - V\frac{R_3(R_2+R_3)}{R_2(R_2+R_3)}, \quad (9.26)$$
and,
$$v_o = v_A + v_A\frac{R_3}{R_2} - V\frac{R_3}{R_2} \quad (9.27)$$

The value of the output voltage given by Expression 9.27 represents the lower limit, A_n, of the output wave, that is, when the voltage v_A reaches the voltage value necessary for the diode D_1 conduction, the output voltage v_o reaches the lower amplitude limit A_n of the oscillator output wave. Thus, the lower limit is given by:
$$A_n = -v_D\left(1 + \frac{R_3}{R_2}\right) - V\left(\frac{R_3}{R_2}\right). \quad (9.28)$$

Figure 9.18 illustrates the behavior of the input and output voltages and the wave with inferior amplitude bounded by the circuit.

Since the output voltage is given by Formula 9.28, the input voltage can be obtained by dividing the output voltage by the gain of the limiter. By increasing the input voltage beyond this value, there is an incremental gain in relation to the input and output voltages. This happens because the increase

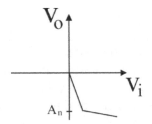

Figure 9.18 Relationship between the input voltage and the output voltage for the positive half-cycle of the input wave.

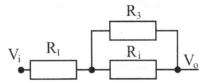

Figure 9.19 Transfer characteristic of the limiter circuit considering only the lower boundary.

of the input voltage causes an increase in the current that passes through D_1. Since the voltage at X point does not change, the current must flow through the resistor R_3 and the circuit can be represented as shown in Figure 9.19.

The incremental gain is given by

$$\frac{v_o}{v_i} = \frac{-R_i \parallel R_3}{R_1}. \tag{9.29}$$

Figure 9.19 illustrates the transfer characteristic considering only the lower boundary. Note that there is a slope in the transfer characteristic, which can be decreased by choosing a low value for R_3.

Case 3: upper limit

The upper limit is found by analyzing the behavior of the limiting circuit when considering the negative half-cycle of the input voltage.

In this case, according to Formula 9.21, the as v_i becomes negative, v_o becomes increasingly positive and, consequently, the voltage v_Y obtained at the Y point of the circuit of Figure 9.16 becomes more and more positive. The voltage in Y point is

$$v_Y = \frac{R_4}{R_4 + R_5}(-V) + \frac{R_5}{R_4 + R_5}v_o. \tag{9.30}$$

That is,
$$v_o \left(\frac{R_5}{R_4 + R_5}\right) = v_Y + V \left(\frac{R_4}{R_4 + R_5}\right). \tag{9.31}$$

And,
$$v_o = v_Y \left(\frac{R_4 + R_5}{R_5}\right) + V \frac{R_4(R_4 + R_5)}{R_5(R_4 + R_5)}. \tag{9.32}$$

Consequently, the output voltage is given by:
$$v_o = v_Y \left(1 + \frac{R_4}{R_5}\right) + V \frac{R_4}{R_5}. \tag{9.33}$$

Considering,
$$v_o = A_p, \tag{9.34}$$

in which A_p is the upper limit of the output, and that the voltage at the Y point is at least equal to the voltage in the diode D_2, to activate its conduction, that is,
$$v_Y = v_D. \tag{9.35}$$

Thus,
$$A_{ps} = V \frac{R_4}{R_5} + v_D \left(1 + \frac{R_4}{R_5}\right). \tag{9.36}$$

Figure 9.20 illustrates the behavior of the input and output voltages and the wave with upper amplitude bounded by the circuit.

Similarly to Case 2, as the input voltage becomes increasingly negative, there is an additional current that must flow through R_4. In this case, the circuit can be visualized as illustrated in Figure 9.22 and incremental gain is given by
$$\frac{v_o}{v_i} = -\left(\frac{R_i \parallel R_4}{R_1}\right). \tag{9.37}$$

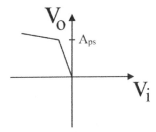

Figure 9.20 Relationship between the input voltage and the output voltage for negative half-cycle of the input wave.

140 Oscillators

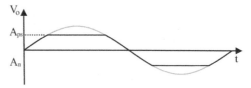

Figure 9.21 Comparison between the sine wave and its output after the limiting circuit.

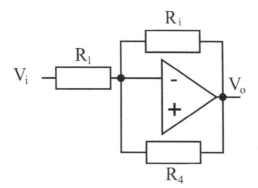

Figure 9.22 Resulting circuit considering the negative half-cycle of the input wave.

General Behavior of the Limit Circuit

Considering the three cases analyzed above, the output of the limiter or non-linear loop circuit should limit the output of the oscillator as shown in Figure 9.21. In addition, the graph resulting from the general transfer feature is shown in Figure 9.23.

9.5 The Wien Oscillator

The Wien Oscillator is one of the most popular linear oscillators in electronics. Its implementation allows the generation of a sinusoidal signal of low or medium frequency and with reduced harmonic distortions.

Figure 9.24 illustrates the electrical circuit of the Wien oscillator. In this configuration, the oscillator does not have a nonlinear control loop. As seen in the figure, this oscillator is implemented from an operational amplifier, two RC meshes, one in series and one in parallel, performing a positive feedback and two resistors, R_1 and R_2 providing the negative feedback (Forouhar Farzaneh, 2018).

For a better analysis of the circuit, consider the following analysis to obtain the closed-loop gain and the loop gain of the circuit.

9.5 The Wien Oscillator 141

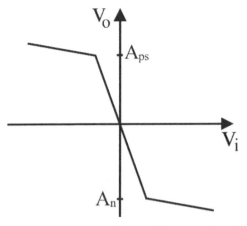

Figure 9.23 General relationship between input and output voltage of the limiting circuit.

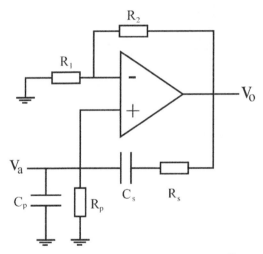

Figure 9.24 Electrical circuit of the Wien oscillator.

The gain of the amplifier is obtained by:

$$\frac{0 - v_a}{R_1} = \frac{v_a - v_o}{R_2}. \tag{9.38}$$

Thus,

$$\frac{v_o}{v_a} = 1 + \frac{R_2}{R_1}. \tag{9.39}$$

142 Oscillators

For the positive feedback loop gain analysis, consider:

$$\frac{0 - v_a}{Z_p} = \frac{v_a - v_o}{Z_s}. \tag{9.40}$$

That is,

$$\frac{v_a(s)}{v_o(s)} = \frac{Z_p}{Z_p + Z_s}. \tag{9.41}$$

Once that the impedance of the mesh in parallel is given by:

$$Z_p = \frac{R_p \times \frac{1}{sC_p}}{R_p + \frac{1}{sC_p}} = \frac{R_p}{sC_pR_p + 1}, \tag{9.42}$$

with $s = \sigma + j\omega$, in which σ and $j\omega$ consist respectively of the real and imaginary part of the complex number.

The impedance of the series mesh is given by:

$$Z_s = R_s + \frac{1}{sC_s} = \frac{sR_sC_p + 1}{sC_s}. \tag{9.43}$$

That is,

$$\frac{v_a(s)}{v_o(s)} = \frac{\frac{R_p}{sC_pR_s+1}}{\frac{R_s}{sC_pR_p+1} + \frac{sC_pR_s+1}{sC_s}}. \tag{9.44}$$

$$\frac{v_a(s)}{v_o(s)} = \frac{\frac{R_p}{sC_pR_s+1}}{\frac{sC_sR_p + s^2 C_p C_s R_p R_s + sC_p R_p + sC_s R_s + 1}{sC_s(sC_pR_s+1)}}. \tag{9.45}$$

$$\frac{v_a(s)}{v_o(s)} = \frac{R_p s C_p}{s^2(C_p C_s R_p R_s) + s(C_s R_p + C_p R_p + C_s R_s) + 1}. \tag{9.46}$$

With $C_p = C_s = C$ and $R_p = R_s = R$ and dividing the numerator and denominator by $\frac{1}{sCR}$,

$$\frac{v_a(s)}{v_o(s)} = \frac{1}{sCR + 3 + 1/sCR}. \tag{9.47}$$

From the transfer functions, we obtain the mesh gain H(s), given by

$$H(s) = \left[1 + \frac{R_2}{R_1}\right]\left[\frac{Z_p}{Z_s + Z_p}\right]. \tag{9.48}$$

9.5 The Wien Oscillator

That is,

$$H(s) = \frac{1 + R_2/R_1}{sCR + 3 + 1/sCR}. \tag{9.49}$$

According to the Barkhausen criterion, the magnitude of the mesh gain must be equal to unity and its phase must be zero. In addition, to have oscillations with constant amplitude at a given frequency the roots of the characteristic equation must be located on the imaginary axis. Thus, the real part is zero and:

$$s = \sigma + j\omega = 0 + j\omega = j\omega. \tag{9.50}$$

Isolating and equating to zero the imaginary term of Formula 9.49:

$$sCR + \frac{1}{sCR} = 0. \tag{9.51}$$

By multiplying both terms by $j\omega CR$ and with $j = -1$, the natural frequency of the oscillation can be found, called the fundamental frequency, ω_o, given by:

$$\omega_o = \frac{1}{CR}. \tag{9.52}$$

In Hertz, the fundamental frequency is given by:

$$\omega_o = \frac{1}{2\pi CR}. \tag{9.53}$$

By Figure 9.24 and by the closed-loop gain of the circuit, it is observed that the Wien oscillator consists of an amplifier connected to a non-inverting configuration. Given the condition established by the Barkhausen criterion of magnitude of the closed-loop gain to be unitary, it is possible to obtain the relation between the resistors of the non-inverting configuration as follows,

$$H(s) = \frac{1 + R_2/R_1}{3} = 1. \tag{9.54}$$

Thus,

$$R_2/R_1 = 2. \tag{9.55}$$

The relation expressed in Formula 9.55 states that for closed-loop gain to be unitary, R_2 should be double that R_1. With this result, for starting oscillations to start in the Wien oscillator, it is necessary that the value of R_2 be slightly greater than twice the value of R_1. In this situation, the mesh gain is greater than unity and the roots of the characteristic equation will be to the right of the half-plane of frequency (Smith A. S., 2004).

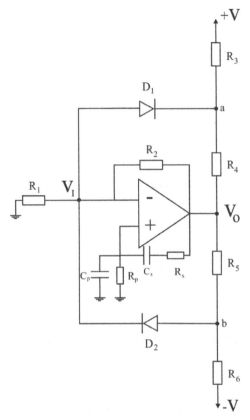

Figure 9.25 Electrical circuit of the Wien oscillator with limiting circuit.

In practice, as mentioned, the oscillators require a nonlinear loop to control the amplitude of the periodic output wave of the circuit and maintain its oscillation at the fundamental frequency.

Figure 9.25 illustrates the Wien oscillator with the non-linear mesh shown in Figure 9.16 for amplitude control.

Using this circuit as a limiter, to control the positive peak of the Wien oscillator output wave amplitude, the voltage at point B must be greater than the voltage v_I, causing the diode D_2 to be conductive.

Using the Formula 9.36 and knowing that

$$v_B = v_I + v_{D2}. \tag{9.56}$$

The positive limit is then found by doing

$$A_{ps} = V\frac{R_5}{R_6} + (v_I + v_{D2})\left(1 + \frac{R_5}{R_6}\right). \quad (9.57)$$

Likewise, the negative peak of the sine wave is obtained by causing the diode D_1 to enter in conduction, with the voltage at point A and the voltage v_I relating as follows

$$v_A = v_I - v_{D1}. \quad (9.58)$$

Thus, the negative peak voltage is given by

$$A_n = -(v_I - v_{D1})\left(1 + \frac{R_4}{R_3}\right) - V\left(\frac{R_4}{R_3}\right). \quad (9.59)$$

Another configuration for the Wien oscillator is shown in Figure 9.26. In this case, the potentiometer can be used to initiate oscillations and control the amplitude of the output sinusoidal signal.

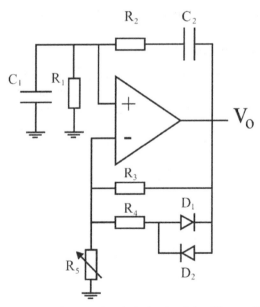

Figure 9.26 Another setting for the electrical circuit of the Wien oscillator with limiting circuit.

9.6 LC Oscillators

The Wien oscillator has good performance in applications requiring low frequencies. However, this oscillator is not suitable for generating signals at high frequencies.

The LC oscillators are an excellent alternative for applications requiring high-frequency sinusoidal signals. These oscillators are easily deployable and can provide output signal frequency ranging from 1 MHz to hundreds of megahertz.

The frequency provided by an operational amplifier is below the range of signal frequency generated by LC oscillators, thus they must be implemented with the use of bipolar junction transistors or FET as amplifier. This type of oscillator receives this designation by using inductive and capacitive circuits to generate the feedback process. The feedback circuit is called the LC tank. Thus, with the use of an amplifier and LC resonant tank, the LC oscillators feedback a signal with phase and amplitude suitable to output a high-frequency sine wave signal.

The following are two of the most popular LC oscillators: Hartley oscillator and Colpitts oscillator (Forouhar Farzaneh, 2018).

9.6.1 The Hartley Oscillator

The Hartley oscillator is a circuit widely used in several applications in electronics. The circuit was invented in 1915 by American engineer Ralph Hartley (1888–1970). One configuration of this circuit is illustrated in Figure 9.27.

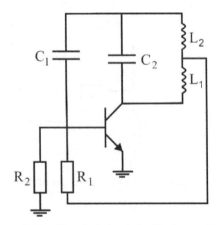

Figure 9.27 Electrical circuit for Hartley oscillator.

9.6 LC Oscillators

The frequency at which the Hartley oscillator will oscillate is found by means of the resonance condition, that is, when the capacitive reactance, X_c, is equal to the inductive reactance, X_L, that is

$$X_c = X_L. \tag{9.60}$$

Therefore,

$$\frac{1}{2\pi f C} = 2\pi f L. \tag{9.61}$$

The oscillation frequency f_H of the Hartlay circuit is given by

$$f_H = \frac{1}{2\pi \sqrt{C L_{eq}}}, \tag{9.62}$$

in which,

$$L_{eq} = L_1 + L_2. \tag{9.63}$$

The principle of operation of the Hartley oscillator is based on a phase inversion of induced current between inductors. Initially, as the circuit illustrated in Figure 9.27 is implemented with the transistor operating as an amplifier, it is necessary considering that the amplifier is operating in a saturation region and, therefore, is in conduction.

The LC tank has the objective to provide the signal feedback in the oscillator. After the saturation, the inductor L_1 induces a current in L_2. Since the inductors have inverted polarities, the current induced will also be an inverted phase (180°). The current with an inverted phase is conducted to the base of the transistor in order to maintain the oscillations. Observe that the transistor inverts the phase of the amplified signal (output signal). Therefore it is necessary that its excitation is done using a signal with an inverted phase, which is guaranteed by inversion of the coils.

The phase inversion process causes the transistor to enter the cut, causing the inductor L_1 to charge again, which induces an inverted current in L_2. This cycle continues and the result is the variation between the states of saturation and cut of the transistor, at a frequency given by the values of the components of the LC circuit.

The variation between saturation and cut happens based on the process of loading and unloading the inductors. The instantaneous current, $i_L(t)$, of the inductor, has a response in the exponential form given by

$$i_L(t) = I_{max}(1 - e^{t/\tau}), \tag{9.64}$$

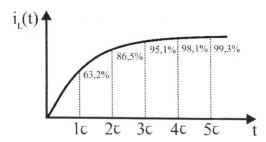

Figure 9.28 Inductor charging rate with time constant.

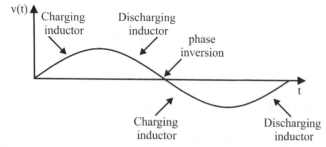

Figure 9.29 Relationship between oscillator output signal and charge, discharge and phase reversal behavior.

in which, I_{max} is the maximum current in steady-state and τ is the time constant. Figure 9.28 illustrates the behavior of the inductor loading.

Considering the state of variation between the cut and saturation states, a sinusoidal output is obtained, as shown in Figure 9.29, in which the first half-cycle of the sine wave is obtained by loading and unloading the inductor L_1, while the second half-cycle is also obtained by the charge and discharge process of the inductor, but with the phase inversion given by the inverted polarity of the inductors.

As mentioned, for the oscillations in an oscillator circuit to start, it is necessary to make the closed-loop gain larger than unity. Thus, consider that the amplifier gain is given by A_H and the feedback gain is given by F_H. In this way, it is

$$A_H F_H > 1. \tag{9.65}$$

In the Harley oscillator, the feedback gain is given through the voltage divider between the inductors. In this oscillator, the output voltage is obtained in the inductor L_1 and the feedback voltage appears on the inductor L_2. The feedback gain is then given by

$$F_H = \frac{L_2}{L_1}. \qquad (9.66)$$

Thus, for the oscillations to start, the gain of the amplifier must be

$$A_H > \frac{L_1}{L_2}. \qquad (9.67)$$

9.6.2 The Colpitts Oscillator

Another very widespread circuit in electronics is the Colpitts oscillator, invented in 1918 by American engineer Edwin H. Colpitts (1872–1949). This oscillator has an operating principle very similar to that of the Hartley oscillator and it is suitable for use at high frequencies.

Figure 9.30 illustrates the electronic circuitry of the Colpitts oscillator. The main difference between this and the Hartley oscillator is that Colpitts provide its resonant tank through a capacitive bypass, rather than an inductive bypass, as in the Hartley oscillator.

The oscillation frequency is obtained from the inductor L and the equivalent capacitor, C_{eq}, given by

$$C_{eq} = \frac{C_1 C_2}{C_1 + C_2}. \qquad (9.68)$$

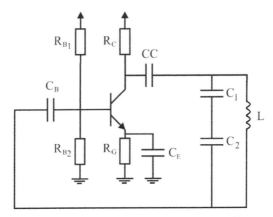

Figure 9.30 Electrical circuit for Colpitts oscillator.

150 Oscillators

Thus, the angular frequency of oscillation w_c of the Colpitts circuit is then given by

$$w_c = \frac{1}{\sqrt{LC_{eq}}}. \tag{9.69}$$

In this oscillator, the resonant circuit, through the capacitive divider, is generating feedback in the base of the transistor, generating the oscillation. The mesh current flows through the capacitor C_1 and the feedback voltage causes the transistor base to sustain the oscillations. The transistor is conducting in the positive half-cycle of the signal and close to the negative half-cycle cut-off. The capacitive divider provides the phase shift of the feedback signal, required for the oscillations to be maintained.

In the Colpitts oscillator, the output voltage appears at the capacitor C_1 and the feedback voltage is at the capacitor C_2. Thus, the feedback gain, F_c, is given by

$$F_c = \frac{C_1}{C_2}. \tag{9.70}$$

And, for the oscillations to start, the gain of the amplifier A_c must be

$$A_c > \frac{C_2}{C_1}. \tag{9.71}$$

9.6.3 The Armstrong Oscillator

The Armstrong oscillator, invented in 1912 by US engineer Edwin Armstrong (1890–1954), is illustrated in Figure 9.31. This oscillator uses a transformer whose second winding feeds back to the transistor base, keeping as output the oscillating signal generated by the resonant tank LC, driven by the resistor collector. In the transformer, there is a 180° phase detachment, causing the circuit loop offset to be zero.

9.7 The Mixer Circuit

Mixers are circuits controlled by an external source and are intended to perform a frequency conversion. They are built by three doors: a radio frequency (RF) port, for obtaining the high-frequency input signal, that is, the signal whose frequency is to be converted by the mixer; a gate to the local oscillator (OL), which is in which the external source used to convert the signal to the new frequency; an intermediate frequency port (IF); which is in which the signal is obtained with a lower frequency component than the input signal or

9.7 The Mixer Circuit 151

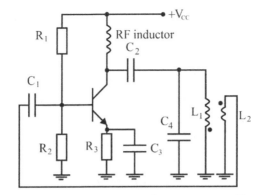

Figure 9.31 Electrical circuit for Armstrong oscillator.

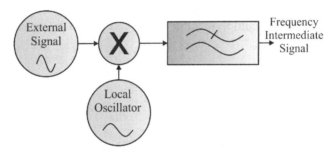

Figure 9.32 General block diagram of a mixer circuit.

signals with frequency components given by the addition and subtraction of the frequencies used by the input signal and the external source.

In the area of communications circuits, they are used in a wide range of applications, such as in modulators to enable the translation of power spectrum, frequency doublers, detectors and phase comparators, demodulators, among other applications.

The basic block diagram of a mixer is shown in Figure 9.32. The mixer has the objective of multiplying two analog signals in real-time so that the output is the sum of signals in frequencies given by the sum and differences in the frequencies of the input signals in the mixer.

Mixers can be classified as passive mixers or active mixers. The classification is given depending on the components that form the mixer circuit. Passives are implemented with diodes and transformers while passive mixers are made with transistors.

The following circuits are described: frequency converter, quadratic mixers, mixers with proportional and quadratic response, passive and active mixers. (Lee, 1998) (Jr, 2005) (Lu, 1999) (Forouhar Farzaneh, 2018) (Taub, 1965)

9.7.1 Mixer as Frequency Converter

The mixer as a frequency converter has the purpose of changing the frequency of one signal to another frequency. This process can be used, for example, in Communications, to change the carrier frequency of a modulated signal.

Let the signal modulated in amplitude, $s(t)$, given by

$$s(t) = m(t)cos(\omega_c t), \tag{9.72}$$

in which $m(t)$ is the message signal and ω_c is the carrier frequency.

The mixer makes it possible to change the carrier frequency, ω_c, for a frequency ω_I, by multiplying the signal modulated by a frequency signal given by

$$\omega_x = \omega_c + \omega_I, \tag{9.73}$$

for upward conversion, or,

$$\omega_x = \omega_c - \omega_I, \tag{9.74}$$

for downward conversion.

Thus,

$$s'(t) = m(t)cos(\omega_c t).cos(\omega_x t). \tag{9.75}$$

Using the trigonometric relations,

$$s'(t) = \frac{m(t)}{2}[cos((\omega_c + \omega_x)t) + cos((\omega_c - \omega_x)t)]. \tag{9.76}$$

Thus, if upward conversion is desired,

$$s'(t) = \frac{m(t)}{2}[cos((2\omega_c + \omega_I)t) + cos(\omega_I)t]. \tag{9.77}$$

Or, if downward conversion is desired,

$$s'(t) = \frac{m(t)}{2}[cos((2\omega_c - \omega_I)t) + cos(\omega_I)t]. \tag{9.78}$$

Considering the results obtained in Expressions 9.77 and 9.78 it is possible to observe that if the new frequency ω_I is greater or equal to the frequency of the message signal, B Hz, there will be no overlapping of the spectra. In addition, the use of a low pass filter, with center frequency in ω_I allow to

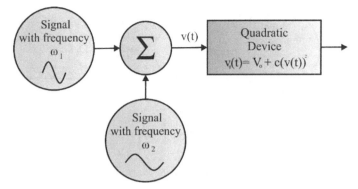

Figure 9.33 Block diagram of a quadratic mixer circuit.

obtain in the output the signal carrier of the message signal, that is,

$$s''(t) = \frac{m(t)}{2}\cos(\omega_I)t. \tag{9.79}$$

9.7.2 Quadratic Mixers

Quadratic mixers are so named because they use devices with a quadratic response. This type of mixer has the objective of obtaining in its output, signals with frequency components given by adding and subtracting the frequencies of the input signal and the oscillator signal.

To do this, as shown in Figure 9.33, the angular frequencies signals ω_1 and ω_2 are summed and input to a quadratic device.

The output of the mixer is the voltage $v_g(t)$ given by

$$v_g(t) = V_o + c(V_1\cos(\omega_1 t) + V_2\cos(\omega_2 t))^2. \tag{9.80}$$

That is,

$$v_g(t) = V_o + c(V_1^2\cos^2(\omega_1 t) + 2V_1V_2\cos(\omega_1 t)\cos(\omega_2 t) + V_2^2\cos^2(\omega_2 t)). \tag{9.81}$$

Using the trigonometric relations,

$$\begin{aligned}v_g(t) = {}& V_o + 0,5cV_1^2 + 0,5cV_1^2\cos(2\omega_1 t) + 0,5cV_2^2 \\ & + 0,5cV_2^2\cos(2\omega_2 t) + V_1V_2\cos((\omega_1 - \omega_2)t) \\ & + V_1V_2\cos((\omega_1 + \omega_2)t).\end{aligned} \tag{9.82}$$

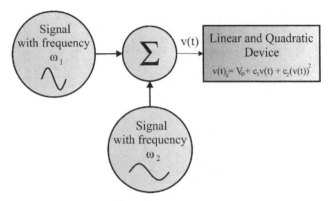

Figure 9.34 Block diagram of a linear and quadratic mixer circuit.

Thus, signals with desired frequency components are obtained by filtering $v_g(t)$, resulting in

$$v_g(t) = V_1 V_2 cos((\omega_1 - \omega_2)t) + V_1 V_2 cos((\omega_1 + \omega_2)t). \quad (9.83)$$

9.7.3 Mixers with Proportional and Quadratic Response

Now consider that the mixer is formed by a device with proportional and quadratic response. That is, as shown in Figure 9.34, the signal resulting from the sum of the signals with frequencies ω_1 and ω_2 is input from a circuit whose response is linear and quadrant.

The output of the mixer is the voltage $v_g(t)$ given by

$$\begin{aligned}v_g(t) = V_o &+ c_1(V_1 cos(\omega_1 t) + V_2 cos(\omega_2 t)) \\ &+ c_2(V_1 cos(\omega_1 t) + V_2 cos(\omega_2 t))^2.\end{aligned} \quad (9.84)$$

Using the trigonometric relations,

$$\begin{aligned}v_g(t) = V_o &+ 0,5c_1 V_1^2 + 0,5c_1 V_1^2 cos(2\omega_1 t) \\ &+ 0,5c_2 V_2^2 + 0,5c_2 V_2^2 cos(2\omega_2 t) + c_1 V_1 cos(\omega_1 t) \\ &+ c_1 V_2 cos(\omega_2 t) + c_2 V_1 V_2 cos((\omega_1 + \omega_2)t) \\ &+ c_2 V_1 V_2 cos((\omega_1 - \omega_2)t).\end{aligned} \quad (9.85)$$

Faced with the result of Expression 9.85, signals with desired frequency components are obtained bypassing $v_g(t)$ through a filter with center frequency given by the desired frequencies. Thus,

$$\begin{aligned}v_g(t) &= c_2 V_1 V_2 cos((\omega_1 + \omega_2)t) \\ &+ c_2 V_1 V_2 cos((\omega_1 - \omega_2)t).\end{aligned} \qquad (9.86)$$

The mixer circuit presents the same behavior when it is considered that the signals pass through generic non-linear devices, that is, in order larger than the second. However, the higher the order of non-linear devices, the more unwanted frequency components appear. These frequency components hamper the filtering process, which is further aggravated when the entrances are considered to be non-sinusoidal. In this way, a mixer should generate as few undesirable components as possible.

9.7.4 Passive Mixers

A passive mixer is implemented using a diode and can be configured in three different ways: mixer with a single diode, balanced mixer, and double-balanced mixer.

Single Mixer:

The general concept of the single mixer is shown in Figure 9.35 and its electronic circuit is illustrated in Figure 9.36. The output of the mixer is then the voltage v_g, obtained under the resistor R.

Since the current in the circuit is the current of the diode, i_D, the voltage in the resistor is given by

$$v_g = R i_D. \qquad (9.87)$$

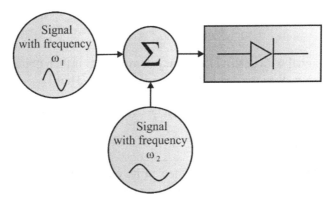

Figure 9.35 Block diagram of a mixer circuit with a single diode.

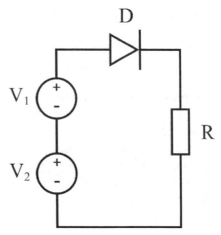

Figure 9.36 Electrical circuit of a mixer with a diode.

Considering the linear and quadratic model,

$$v_g = R(c_1 v_D + c_2 v_D^2), \qquad (9.88)$$

in which the voltage in the diode v_D, is given by

$$v_D = V_1 + V_2. \qquad (9.89)$$

Thus, the voltage v_g is given by

$$\begin{aligned}v_g = R[&0,5c_1 V_1^2 + 0,5c_1 V_1^2 \cos(2w_1 t) + 0,5c_2 V_2^2 \\ &+ 0,5c_2 V_2^2 \cos(2w_2 t) + c_1 V_1 \cos(w_1 t) \\ &+ c_1 V_2 \cos(w_2 t) + c_2 V_1 V_2 \cos((w_1 + w_2)t) \\ &+ c_2 V_1 V_2 \cos((w_1 - w_2)t)]. \end{aligned} \qquad (9.90)$$

The signal with desired frequency components is then obtained bypassing the voltage v_g through a suitably configured filter. Thus,

$$v_g = R[c_2 V_1 V_2 \cos((w_1 + w_2)t) + c_2 V_1 V_2 \cos((w_1 - w_2)t)]. \qquad (9.91)$$

Balanced Mixer:

The mixer is named balanced when in its circuit there are two diodes. The balanced mixer circuit is shown in Figure 9.37.

9.7 The Mixer Circuit

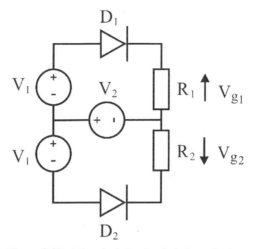

Figure 9.37 Electrical circuit of a balanced mixer.

Observing the circuit,

$$v_{g1} = R[0,5c_1V_1^2 + 0,5c_1V_1^2\cos(2\omega_1 t) + 0,5c_2V_2^2 \\ + 0,5c_2V_2^2\cos(2\omega_2 t) + c_1V_1\cos(\omega_1 t) + c_1V_2\cos(\omega_2 t) \\ + c_2V_1V_2\cos((\omega_1+\omega_2)t) + c_2V_1V_2\cos((\omega_1-\omega_2)t)], \quad (9.92)$$

and,

$$v_{g2} = R[0,5c_1V_1^2 + 0,5c_1V_1^2\cos(2\omega_1 t) + 0,5c_2V_2^2 \\ + 0,5c_2V_2^2\cos(2\omega_2 t) - c_1V_1\cos(\omega_1 t) + c_1V_2\cos(\omega_2 t) \\ - c_2V_1V_2\cos((\omega_1+\omega_2)t) - c_2V_1V_2\cos((\omega_1-\omega_2)t)]. \quad (9.93)$$

The output voltage of the mixer is given by

$$v_g = v_{g1} - v_{g2}. \quad (9.94)$$

Therefore,

$$v_g = R[2c_1V_1\cos(\omega_1 t) + c_2V_1V_2\cos((\omega_1+\omega_2)t) + c_2V_1V_2\cos((\omega_1-\omega_2)t)]. \quad (9.95)$$

As in the case of the single diode mixer, the balanced mixer obtains the signal with desired components through the passage v_g of filters centered on the desired frequency component. Therefore, the output signal is given by

$$v_g = R[c_2V_1V_2\cos((\omega_1+\omega_2)t) \\ + c_2V_1V_2\cos((\omega_1-\omega_2)t)]. \quad (9.96)$$

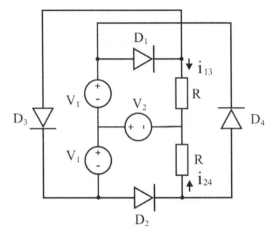

Figure 9.38 Electrical circuit of a double-balanced mixer.

Double-Balanced Mixer:

The double-balanced mixer is easy to implement and widely used in communications systems for signal transmission and reception, as well as in applications involving signal processing.

A practical way to implement this type of mixer is illustrated in Figure 9.38, in which it is observed that the circuit is formed by two transformers connected to a 4-diode bridge.

In front of the circuit of Figure 9.38, it is observed that the output voltage of the mixer v_g is given by

$$v_g = v_{g1} - v_{g2}, \tag{9.97}$$

with,

$$v_{g1} = Ri_{13}, \tag{9.98}$$

and,

$$v_{g2} = Ri_{24}. \tag{9.99}$$

Thus,

$$v_g = R(i_{13} - i_{24}), \tag{9.100}$$

with $i_{13} = i_{D1} - i_{D3}$ and $i_{24} = i_{D2} - i_{D4}$.

Based on the above expressions, the voltage v_g is given by

$$v_g = R(i_{D1} - i_{D3} - i_{D2} + i_{D4}). \tag{9.101}$$

9.7 The Mixer Circuit

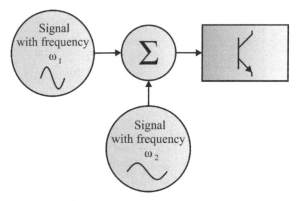

Figure 9.39 Block diagram of an active mixer.

Using the relationship between voltage and current of the diode for a mixer with linear and quadratic response given by Formula 9.87 and Expression 9.88, the expression for the final tension of the mixer is given by

$$v_g = 2R[c_2 V_1 V_2 cos((\omega_1 + \omega_2)t) + c_2 V_1 V_2 cos((\omega_1 - \omega_2)t)]. \tag{9.102}$$

9.7.5 Active Mixers

The active mixers are those formed by transistors, as illustrated by Figure 9.39, and can be of the simple, balanced and doubly balanced type. An example of an active mixer is shown in Figure 9.40.

In this circuit, it is observed that the base-emitter voltage, v_{BE}, is given by

$$v_{BE} = v_1 + v_2 = V_1 cos(\omega_1 t) + V_2 cos(\omega_2 t). \tag{9.103}$$

In addition, the output voltage of the mixer is the voltage obtained on the resistor R and can be found by the following relation

$$v_g = R i_C, \tag{9.104}$$

in which i_C is the collector current of the transistor. It is also observed that the current in the collector has an inverse direction of the current of the resistor, that is,

$$i_C = -i_R. \tag{9.105}$$

Considering the linear and quadratic relationship of the collector current expressed by

$$i_C \approx i_R + c_1 v_{BE} + c_2 v_{BE}^2. \tag{9.106}$$

160 Oscillators

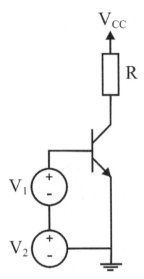

Figure 9.40 Electrical circuit of an active mixer.

Thus,

$$v_g = R[i_R + 0,5c_2V_1^2 + 0,5c_2V_1^2\cos(2w_1 t) + 0,5c_2V_2^2 \\ + 0,5c_2V_2^2\cos(2w_2 t) + c_1V_1\cos(w_1 t) + c_1V_2\cos(w_2 t) \\ + 0,5c_2V_1V_2\cos((w_1+w_2)t) + 0,5c_2V_1V_2\cos((w_1-w_2)t)]. \tag{9.107}$$

Obtaining the output voltage from the mixer given by Expression 9.107, results in signals having components of undesirable frequencies, that is, frequencies w_1, w_2, $2w_1$ and $2w_2$ which can be eliminated by filtering.

9.8 Voltage Control Oscillator

The Voltage Control Oscillator (VCO) is one of the most popular circuits in electronics and widely used in the field of communications circuits, especially in systems using angular modulation and Phase-Locked Loop (PLL).

The VCO consists of an oscillator whose oscillation frequency of the output signal is controlled by means of an input voltage. In other words, the VCO circuit allows obtaining an output signal with frequency that changes within a range determined by certain components of the circuit, from a voltage variation at the input of the circuit.

9.8 Voltage Control Oscillator

The VCO can be linear (or harmonic) or relaxation type. The VCO as a linear oscillator produces a sinusoidal signal at its output, while the VCO of the relaxation type generates at its output a rectangular, triangular wave or sawtooth signal.

The oscillation frequency, $\omega(t)$, of the output signal of the VCO is given by

$$\omega(t) = \omega_o + K v_e(t), \quad (9.108)$$

in which ω_o is a mean oscillation frequency, K is the gain of the amplifier, whose unit is hertz per volt, and $v_e(t)$ is the input voltage which controls the frequency variation.

The phase of the output signal is given by

$$\theta(t) = \int_{-\infty}^{t} f(\tau) d\tau. \quad (9.109)$$

In addition to the parameters presented in the Expressions 9.108 and 9.109, the VCO has other parameters, such as:

- Harmonic Suppression: informs the intensity ratio between the fundamental frequency component and its harmonics.
- Tuning Sensitivity: Specifies the sensitivity in changing the frequency of the output signal when the input voltage changes. The unit of this parameter is hertz per volt (Hz/V).
- Tuning Range (TR): Available frequency range for the VCO output signal versus the input signal voltage range. It is given by

$$TR = 2 \left(\frac{f_{max} - f_{min}}{f_{max} + f_{min}} \right). \quad (9.110)$$

- Load Pulling: measure the sensitivity of the frequency in relation to the variation of its load.
- Supply Pulling: this parameter is expressed in hertz per volts and measures the sensitivity of the frequency of the output signal versus the variation of the input voltage.
- Spectral purity: is a measure that relates the spectrum of the output signal to the noise generated in the VCO process. It can be represented by means of jitter or in terms of phase noise.

A possible implementation of the relaxation VCO circuit is shown in Figure 9.41. In this circuit, a triangular wave is generated from an integrative mode amplifier and Schmitt Trigger (hysteresis), an input voltage and the positive and negative supply meshes. The triangular wave has a lower amplitude

162 *Oscillators*

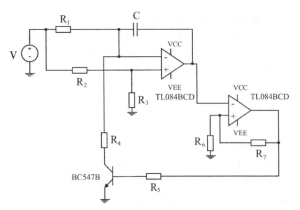

Figure 9.41 VCO circuit.

or, at most, given by the supply voltage of the TL084 CI. The triangular wave is the input signal that feeds the second operational amplifier, which is also in hysteresis configuration, to generate the square wave. The output is then fed back by the transistor BC547, which controls the resulting frequency of the signal by means of the input DC voltage.

10

The Phase-Locked Loop

The Phase-Locked Loop (PLL) is an electronic system used in practically all analog and digital communications systems. The first known application of PLL was in 1932, for synchronism detection in digital radio.

In analog communications, for example, a PLL is used in the synchronous demodulation of amplitude-modulated signals without sending the carrier (AM-DSB-SC) or in the demodulation of amplitude-modulated signals by sending the carrier in transmission systems having low signal-to-noise ratio. In addition, PLL can also be used in the modulation and demodulation of angle modulated signals.

In digital communications, PLL is useful in clock recovery for synchronization of digital transmission, demodulation of FSK signals, among others.

This chapter introduces the PLL circuit, describing all the steps for its circuit, such as the Voltage-Controlled Oscillator (VCO), mesh filter, and phase comparator. In addition, this chapter presents the mathematical modeling of PLL, introduces digital PLL and PLL as a frequency synthesizer.

10.1 General Description of PLL

The operation principle of PLL is based on a negative feedback system, as shown in Figure 10.1. When the loop gain, g_L, is considered to be large, the error signal, $e_a(t)$, given by the difference between the input signal, $v_{ia}(t)$, and the feedback signal, $v_{ar}(t)$, has its value close to zero. Therefore, under these conditions, it is possible to say that the feedback signal tends to follow the input signal (Smith, 2004).

In the case of PLL, the comparison is made between an internal oscillator and the external signal (input signal), in order to trace the phase and frequency of the external signal.

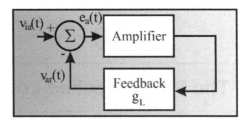

Figure 10.1 Block diagram of a negative feedback system.

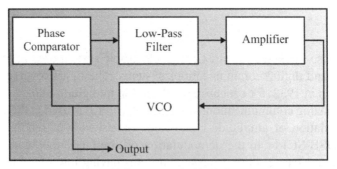

Figure 10.2 PLL basic block diagram.

Thus, differently of the negative feedback system, which compares amplitude, the PLL compares the phase and frequency between an external signal, which may be a sinusoid or digital signal, and an internal signal, represented by the output of the VCO, until the signal internal, generate an output that accompanies the phase and frequency of the external signal.

Figure 10.2 illustrates the basic PLL architecture, in which it is possible to observe that it is formed by four components: the voltage-controlled oscillator (VCO), a low-pass filter, a phase comparator and an amplifier.

The main idea of PLL is to reproduce at its output, a signal that has the same phase and frequency characteristic as the input signal. This process is important in communications because, among other reasons, it, by reproducing the input signal, eliminates possible noise components that it has in the original signal and generates a clean signal, that is, without adverse interference of the communication channel. In addition, the PLL reproduces at its output a more stable signal with less jitter interference, if the mesh filter is designed with a cutoff frequency less than the jitter frequency.

The PLL receives an input signal and, from the VCO, generates an output signal with the same phase and frequency of the input signal. Once the signals have the same phase and frequency, they are in synchronism and the circuit

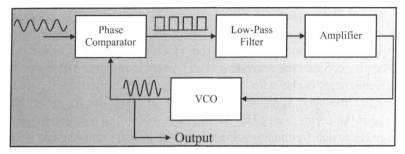

Figure 10.3 PLL detailed block diagram with sine input and output.

reaches the locked state. However, there may still be a possible constant phase difference between the signals.

Figure 10.3 illustrates a more detailed diagram of the PLL. The input signal, in this example, sinusoidal, is compared to a sinusoidal signal generated by the VCO, with frequency initially given from the characteristics of the VCO circuit.

The comparator output, called an error signal, is a pulsed signal, with pulses whose width represents the discrepancy between the signals.

The VCO is an oscillator whose frequency of the output signal is adjusted according to the voltage variation of its input signal. The average voltage generated by the low pass filter from the error signal is input from the VCO and will modify the frequency of the signal generated by it so as to leave it equal to the frequency of the input signal in the PLL.

The entire PLL operation takes place within two frequency ranges, called the capture range and the locking range.

The PLL is capable of reproducing a signal with the same frequency and possibly the same phase if the input signal has a frequency within the PLL capture range. In this case, an error signal is generated at the output of the phase comparator, in which the mean value of which error signal changes the frequency of the output signal of the VCO to the same frequency value of the input signal of the PLL.

It is important to note that the signal reproduced by the VCO must have the same frequency as the input signal in the PLL, but they can have a constant phase difference. Since the phase difference between the signals is constant, the error signal is also constant, causing the input voltage to sustain the new frequency of the VCO.

Despite a possible phase difference, if the signals have the same frequency, the PLL reaches the state of synchronism between the signals, in other words, it is locked.

166   The Phase-Locked Loop

The possible phase difference between the signals, if constant, does not change the state of synchronism of the PLL, since the instantaneous frequency of the output signal of the VCO is proportional to the derivative of the phase. Since the derivative of a constant is zero, the difference between the frequencies is also zero, causing the signals to be in sync.

After the sync state, the PLL is able to track the frequency variation of the input signal, since this variation is within the locking range of sync range. For each frequency value within the locking range, there is only one phase difference and, hence the average voltage obtained on the mesh filter, consisting of the VCO input, which changes the frequency of the VCO to the new frequency of the input signal in PLL.

To better understand the operation of PLL, each of the steps is described in detail below.

10.1.1 Voltage-Controlled Oscillator (VCO)

The Voltage Controlled Oscillator is an oscillator whose frequency of the output signal is controlled by a voltage at its input. In it, there is a linear relationship between frequency and voltage and in PLL, the VCO is used to produce a signal with the same frequency of the input signal.

The VCO generates a sinusoid with a frequency that varies in line with the voltage applied to its input. If there is no external signal applied to the phase detector, the VCO operates in the free-running mode, that is, it generates a signal whose frequency is related to the design of its circuit.

In the case of an external signal, the phase detector will generate an error signal with information of the phase difference between the signals. The error signal is then filtered and the resulting voltage is applied to the VCO, forcing it to change its output signal to a new phase and frequency sinusoid that is closer to those presented by the external signal, thereby reducing the error between the two signals.

For a better understanding of the operation of the VCO in PLL, consider the following two cases.

1st Case: constant phase difference between signals

If the error is constant, that is, if the phase difference between the input signal and the output signal of the VCO is always the same, the value of the average output voltage of the low pass filter will be constant and, consequently, there will be no change in the oscillation frequency of the VCO output signal. This

10.1 General Description of PLL

indicates that the oscillation frequency of the VCO output signal has the same frequency as the input signal.

2nd Case: variable phase difference between signals

In case of a frequency difference between the input signal and the internal signal, the output of the error comparator will be variable, resulting in a variable voltage at the output of the filter. This variable voltage forces the VCO to adjust the frequency of its output signal to accompany the input signal in the PLL. In this case, higher amplitude voltages increase the oscillation frequency of the VCO signal, while voltages with lower amplitude decrease the oscillation frequency of the VCO output signal.

The process continues until the phase difference is sufficiently small or constant and consequently the average voltage generated by the mesh filter is also constant and, thus, not sufficient to force the VCO to generate a signal with a new frequency. In this case, the external and internal signals are synchronized, that is, with the same frequency and the circuit is locked or in synchronism.

Figure 10.4 illustrates the process of convergence of the VCO signal frequency. In the figure, the VCO signal is initially generated at frequency f_o, given by the parameters chosen in its construction. Until to the instant of time t_1 there is no input signal in the PLL and the output signal of the VCO oscillates in its free frequency, that is, f_o.

At the instant of time t_1, a signal with the oscillation frequency f_i is inserted into the PLL input and the PLL reaches its synchronism state at the instant of time t_2.

The time interval $t_1 < t < t_2$ represents the transient state interval of the PLL. The transient state refers to how the output voltage of the filter accompanies the rapid or even abrupt variations of the input signal frequency

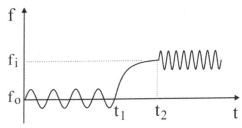

Figure 10.4 The process of convergence of the VCO signal frequency, from f_o to f_i.

Figure 10.5 Two sinusoidal signals, with time or phase difference, representing the input signal in the PLL and the VCO signal.

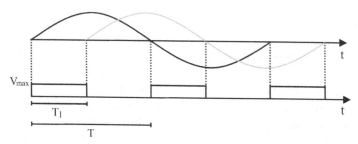

Figure 10.6 Representation of the phase comparator output generation from the comparison between the sinusoid generated by the VCO and the input signal in the PLL.

in the PLL. Under real conditions, the mesh filter and the VCO introduce variable delays and delays causing the PLL response not to be instantaneous.

10.1.2 Phase Comparator

The phase comparator or mixer, whose mathematical formulation is presented in Chapter 9, operates with a phase detector. In it, the signal arriving at the PLL is compared in phase and frequency with the generic internal signal generated by the VCO. The output of the phase detector then consists of an error signal, which is a function of the phase difference between the external and internal signals.

To better understand the operation of the mixer, consider Figure 10.5 which illustrates two sine signals with a given phase difference between them. These signals may be, for example, the voltage lag between the resistor and a capacitor, or the voltage lag between a capacitor and an inductor.

The lag between the signals can be obtained as a function of time, Δt, or as a function of phase angle, $\Delta \theta$. The output of the phase detector consists of a pulse train, in which each pulse occurs in the delay interval between the signals and its duration represents the phase delay. Figure 10.6 shows the output of the phase comparator, in which it is possible to observe that each pulse is located in the lag interval, that is, in which each signal passes through zero.

The pulse train resulting from the phase detector or phase comparator is called the error signal. That is, the longer the duration of the pulses, the greater the discrepancy between the signals, that is, the greater the error between them.

The idea is then, from the error signal, to obtain the average voltage, V_m, which is given by

$$V_m = \frac{T_1}{T} V_{max}, \qquad (10.1)$$

in which V_{max} is the maximum pulse voltage, T is the period of the train of pulses and T_1 is the duration of the pulse.

By Formula 10.1, it is verified that the greater the error between the input signals of the comparator, the greater the average voltage generated. The error signal is filtered, amplified, and its DC component is applied to the VCO input.

10.1.3 Low Pass Filter

The theory about the low-pass filter has already been introduced in this book. Here is a focused on filter operation for use in PLL.

The input signal of the low pass filter is the signal resulting from the comparison performed between the internal signal of the VCO and the input signal of the PLL. This signal, which is the error signal, has the characteristic of being a pulsating signal.

The low pass filter is implemented by a resistor in series with a capacitor, as shown in Figure 10.7.

In the case of use in PLL, as mentioned, the input signal is a pulsating signal, characteristic of the error signal. The purpose of the low-pass filter is then to obtain the mean of the pulsed signal, as illustrated in Figure 10.8.

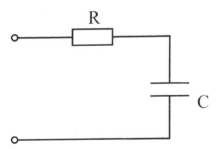

Figure 10.7 Low pass filter circuit.

170 The Phase-Locked Loop

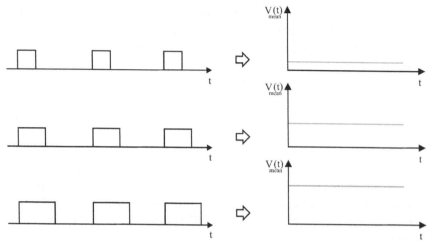

Figure 10.8 Relationship between the error signal and the filter output. For a higher error signal, the higher the average voltage at the filter output will be.

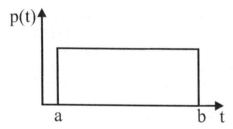

Figure 10.9 Pulse function.

The error signal is a pulse train and to observe the functionality of the low-pass filter in the PLL it is essential to understand the filter response when there is a sequence of step functions, characteristic of the pulse train.

Each pulse, $p(t)$, as shown in Figure 10.9, can be mathematically represented by the step function, $u(t)$, as (Lathi, 2004)

$$p(t) = u(t-b) - u(t-a). \tag{10.2}$$

As the input of a low pass filter, the pulse represents the insertion of a DC source applied suddenly and its response is known as a response to a step. In this way, understanding the operation of the low pass filter in the PLL circuit is to understand the response of an RC circuit when excited by the step function.

10.1 General Description of PLL

Consider a voltage, V_s, applied to the RC circuit. The voltage is divided between the components of the circuit and, applying Kirchofft's law of the currents, (Riedel, 2008).

$$C\frac{dv}{dt} + \frac{v}{R} - \frac{V_s u(t)}{R} = 0, \qquad (10.3)$$

which can be written as,

$$\frac{dv}{dt} + \frac{v}{RC} = \frac{V_s}{RC}u(t). \qquad (10.4)$$

If the capacitor has an initial voltage $v(0) = v_0$, the response of an RC circuit to the application at a DC voltage, to $t > 0$ is

$$v(t) = v_s + (v_o - v_s)e^{(\frac{-t}{\tau})}, \qquad (10.5)$$

in which τ is the time constant of the capacitor.

Figure 10.10 illustrates the charging behavior of the capacitor against DC voltage, considering the capacitor initially charged (Figure 10.10(a)) and capacitor initially discharged (Figure 10.10(b)).

Thus, when a DC voltage of magnitude v_s is applied, the capacitor will charge up to its maximum value, which is given by the applied DC voltage. In addition to the charging curve, it is important to note the capacitor discharge curve, as shown in Figure 10.11. In this case, when removing the voltage source, the capacitor will discharge, from the maximum value v_s until it reaches zero.

The capacitor charging and discharging curves are important for understanding the behavior of the low-pass filter output signal. When a pulse is applied to the filter input, for example, at the instant of time t_a, the capacitor

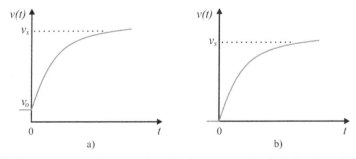

Figure 10.10 Curve illustrating capacitor charging: (a) capacitor initially charged; (b) capacitor initially discharged.

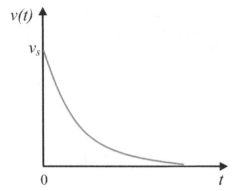
Figure 10.11 Curve illustrating the capacitor discharge.

starts charging, as shown in Figure 10.12(a), and, at the end of the pulse, at the instant of time t_b, the capacitor begins to discharge, as illustrated by Figure 10.12(b).

Considering this behavior for all pulses of the pulse train present in the error signal resulting from the comparison in the PLL circuit between the input signal and the internal signal, there is a loading and unloading sequence of the capacitor, resulting in waveforms, as shown in Figure 10.13.

The time constant, given by Formula 10.6, represents the charge and discharge time of the capacitor.

$$\tau = RC. \tag{10.6}$$

The charge and discharge behavior of the capacitor is in relation to the time constant. In charge of the capacitor, the time constant represents the time the capacitor takes to charge with 63% of the maximum voltage applied. In the discharge phase, the time constant consists of the time the capacitor took to discharge 64% of the maximum voltage, that is, only 36% of the maximum charge voltage remains in the capacitor.

10.1.4 PLL Capture Range

An important parameter to observe in PLL is the capture range, that is, the frequency range at which the VCO is able to initially trace the input signal (external signal) and remain locked.

Figure 10.14 illustrates the behavior of the PLL in relation to the capture range and the introduction of an external signal.

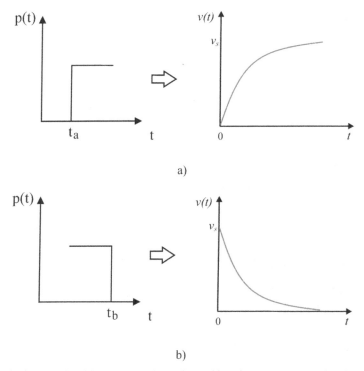

Figure 10.12 Relationship between the pulse, which forms the error signal, with the capacitor charging and discharging curves.

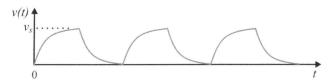

Figure 10.13 Curve representing various capacitor charging and discharging cycles.

The capture range encompasses frequencies around of the central oscillation frequency, f_0, which is the natural oscillation frequency offered by the VCO circuit. It consists of the frequency range between the frequencies f_1 and f_2, and therefore, the VCO is able to track and lock at any external signal frequency that is contained within that frequency range.

Initially, prior to the presence of the input signal, the VCO operates at its central oscillation frequency. After inclusion, at the instant of time t_1, of the input signal with frequency f_i, the oscillation frequency of the VCO

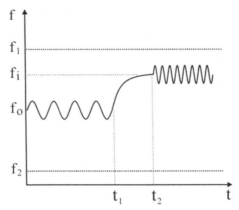

Figure 10.14 PLL sync within capture range, that is, between the frequencies f_1 and f_2.

converges to the frequency of the input signal, since it is within the capture range. In the case in which the frequency of the external signal is outside the capture range, the VCO will produce a signal with the central oscillation frequency, which is the frequency given by the VCO circuit.

10.1.5 PLL Lock Range

The locking range represents another important parameter in the PLL construction. After the VCO tracks the frequency of the input signal, respecting the capture range, the PLL lock range consists of the frequency range that the VCO is able to track the frequency variations of the input signal.

As shown in Figure 10.15, the locking range comprises the interval between F_1, which is the lower limit of the locking range, and F_2 which is the upper limit of the locking range.

Thus, after the VCO performs the frequency tracking, at instant t_2, it is possible to vary the frequency of oscillation of the input signal within the limits of the locking range.

10.2 Mathematical Model of PLL

After analyzing the systems that make up the PLL, this section presents the mathematical model that describes the PLL.

Initially, consider the signal that the input signal, $V_i(t)$, in PLL is given by (Ding, 2010)

$$V_i(t) = V_i \cos(\omega_i t + \theta_i), \qquad (10.7)$$

10.2 Mathematical Model of PLL

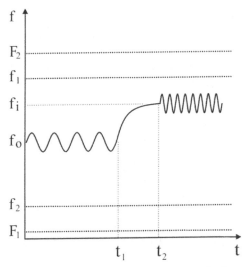

Figure 10.15 PLL sync within lock range, that is, between the frequencies F_1 and F_2.

in which V_i, ω_i and θ_i are, respectively, the amplitude, angular frequency, and phase of the input signal.

Consider also that the signal, $V_o(t)$, initially generated by the VCO is given by

$$V_o(t) = V_0 \cos(\omega_0 t + \theta_0), \tag{10.8}$$

in which V_0, ω_0 and θ_0 are, respectively, the amplitude, angular frequency, and phase of the VCO signal. Its instantaneous frequency, ω_{inst}, is given by

$$\omega_{inst}(t) = \frac{d}{dt}[\omega_0 t + \theta_0]. \tag{10.9}$$

or,

$$\omega_{inst}(t) = \omega_0 + \frac{d\theta}{dt}. \tag{10.10}$$

The signal, $x(t)$, resulting from the phase comparator is given by

$$x(t) = V_i \cos(\omega_i t + \theta_i) V_0 \cos(\omega_0 t + \theta_0). \tag{10.11}$$

After the use of the trigonometric relation between two cosines,

$$x(t) = \frac{V_i V_0}{2}[\cos((\omega_0 + \omega_i)t + \theta_0 + \theta_i) + \cos((\omega_0 - \omega_i)t + \theta_0 - \theta_i)]. \tag{10.12}$$

Now, consider that the angular frequency of central oscillation of the VCO is given by ω_0. The VCO is an oscillator whose frequency of the output signal

176 The Phase-Locked Loop

is controlled to obtain the frequency of the input signal. The frequency control of the VCO is given linearly, from an input voltage, $v(t)$, which is the output voltage of the low pass filter.

Thus, if the input voltage of the VCO is $v(t)$, the instantaneous frequency of the output signal of the VCO is given by

$$w(t) = w_0 + av(t). \tag{10.13}$$

in which the parameter a consists of the VCO control constant. From Expression 10.13

$$\frac{d\theta(t)}{dt} = av(t). \tag{10.14}$$

If $v(t) = 0$, the VCO oscillates at its angular frequency of central oscillation or free frequency w_0. In this case, the loop is stabilized and the input signal is assumed to have the same frequency as the VCO signal. Thus, Expression 10.12 can be rewritten as

$$x(t) = \frac{V_i V_0}{2}[\cos((2w_0)t + \theta_i + \theta_0) + \cos((w_i - w_0)t + \theta_i - \theta_0)]. \tag{10.15}$$

The first frequency summation term of Expression 10.15 represents the second harmonic and is rejected in passing $x(t)$ through the low pass filter. In order that, in the stabilized circuit condition, the voltage $v(t)$ is zero, the feedback gain must produce a 90° phase shift in the VCO frequency to force the second term, of the frequency subtraction, to be zero.

Now consider the locked loop, that is, $w_0 = w_i$. The output of the phase comparator is given by Formula 10.16, assuming that the frequency sum term is suppressed by the filter.

$$x(t) = \frac{V_i V_0}{2} \cos(\Delta\theta(t)), \tag{10.16}$$

in which $\Delta\theta(t) = \theta_i(t) - \theta_0(t)$ is the phase error.

The signal $x(t)$ is input from the low-pass filter, whose output $v_m(t)$ is given by

$$v_m(t) = x(t) * h(t), \tag{10.17}$$

in which $h(t)$ is the impulse response of the low pass filter. Thus, for $t > 0$,

$$v_m(t) = \int_0^t \frac{V_i V_0}{2} \cos(\Delta\theta(x)) h(t-x) dx. \tag{10.18}$$

And,

$$\frac{d\theta(t)}{dt} = \int_0^t \frac{aV_i V_0}{2} \cos(\Delta\theta(x)) h(t-x) dx. \tag{10.19}$$

10.2 Mathematical Model of PLL

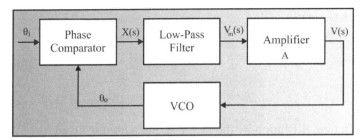

Figure 10.16 Block diagram illustrating PLL in the frequency domain.

10.2.1 Analysis of PLL under Small Signals

To analyze the PLL under small signals, consider Figure 10.16, which illustrates the block diagram of the PLL in the frequency domain.

From Figure 10.16 it is possible to observe that

$$V(s) = AV_m(s). \tag{10.20}$$

in which $V(s)$ and $V_m(s)$ are respectively the Laplace transform of the voltage $v(t)$ and $v_m(t)$, and A is the gain of the amplifier. Moreover, considering the analysis for small signal, that is, in which it is possible to do approximate $\operatorname{sen} x \approx x$, the output signal of the phase comparator, $x(t)$, is given by

$$x(t) = G\Delta\theta. \tag{10.21}$$

or,

$$x(t) = G[\theta_i - \theta_0], \tag{10.22}$$

in which G is the gain of the comparator given by

$$G = \frac{V_i V_o}{2}. \tag{10.23}$$

The output of the low-pass filter, in the frequency domain, $H(s)$, is given by

$$H(s) = \frac{V_m(s)}{X(s)}. \tag{10.24}$$

in which $X(s)$ is the Laplace transform of the signal $x(t)$. In addition, it is known that it is possible to express the VCO signal phase by

$$\theta_0 = \int \Delta\omega_o(t) dt. \tag{10.25}$$

178 The Phase-Locked Loop

Consequently,

$$\frac{d\theta}{dt} = \Delta\omega_o(t) = aV(t). \tag{10.26}$$

$$\theta_o(s) = \frac{1}{s}\Delta\omega_o(s) = \frac{1}{s}aV(t). \tag{10.27}$$

After amplification by the factor A, the output is given by

$$V(s) = AH(s)G(\theta_i - \theta_o). \tag{10.28}$$

Using Formula 10.27 in Formula 10.28,

$$\frac{V(s)}{\theta_i(s)} = \frac{AH(s)G}{1 + \frac{AH(s)aG}{s}}. \tag{10.29}$$

Entering the input frequency by means of the ratio

$$\Delta\omega_i = s\theta_i(s). \tag{10.30}$$

Thus,

$$\frac{V_o(s)}{\Delta\omega_i} = \frac{V_o(s)}{s\theta_i(s)} = \frac{AH(s)G}{s + AH(s)aG}. \tag{10.31}$$

Formula 10.31 is in the form of a closed-loop feedback function, whose magnitude of the mesh gain is given by $AH(s)aG$.

Considering the phase difference of the input signal and the VCO signal, that is,

$$\Delta\theta_i = \theta_i - \theta_o. \tag{10.32}$$

Thus,

$$\Delta\theta_i = \frac{s\theta_i}{s + AH(s)aG}. \tag{10.33}$$

Thus, if it is considered that the magnitude of the mesh gain is large,

$$\Delta\theta_i = \frac{s\theta_i}{AH(s)aG}. \tag{10.34}$$

From Formula 10.34 it is observed that, for a large magnitude of the loop gain, the magnitude of the phase error is reduced to zero with a phase shift 90° of from the input signal.

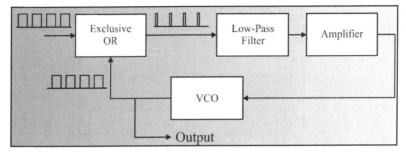

Figure 10.17 Block diagram illustrating the digital PLL.

10.3 The PLL Digital Circuit

The digital PLL operating circuit is shown in Figure 10.17.

Since the error signal between the input signals in the phase comparator will only exist at the time of lag, that is, when a signal and absence of the second signal is in the comparison, the phase comparator is implemented by the logic operation XOR, in which there is an output with logic level 1 when comparing different logical levels, that is 1 and 0 or 0 and 1.

10.4 The PLL as Frequency Synthesizer

An important application of PLL is as a frequency synthesizer. These electronic circuits are used, for example, in superheterodyne receivers.

Superheterodyne receivers are used, for example, to receive AM and FM radio signals. In this type of receiver, incoming signals are translated to an intermediate frequency and then demodulated and this frequency translation is done using PLL as a frequency synthesizer.

The PLL, in this case, aims to provide at its output a signal with a frequency multiple of the frequency of the input signal, that is, it translates the modulated carrier frequency to high or low frequencies.

For this, a digital divider after the VCO is included in the PLL, as shown in Figure 10.18. In this case, the frequency of the signal generated by the VCO must be multiple of the factor M of the divisor, so that after the signal passes the divider, it results in a signal whose frequency is the same as the input signal in the PLL.

In this way, we get signals translated in frequency by a factor M of the divisor. It is important to note that when PLL is used as a frequency synthesizer, its output is also given by the signal generated by the VCO.

180   The Phase-Locked Loop

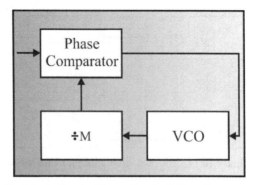

Figure 10.18   Block diagram illustrating the PLL as frequency synthesizer.

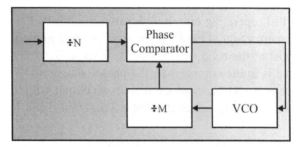

Figure 10.19   Another configuration of PLL as frequency synthesizer.

In frequency synthesizers, the oscillation frequency of the VCO output signal is given by

$$f_{os} = M f_i. \quad (10.35)$$

The PLL configuration may also include a digital divider for the input signal. The digital divider is then positioned before the phase comparator, as shown in Figure 10.19 and thus it is possible to perform the multiplication by fractional values. In this case, the signals are synchronized when

$$\frac{f_{os}}{M} = \frac{f_i}{N}. \quad (10.36)$$

And, consequently, the oscillation frequency of the VCO is given by

$$f_{os} = \frac{M}{N} f_i. \quad (10.37)$$

11

Continuous Wave Modulation

A communications system is intended to transmit a signal carrying information from its source to its destination. In its most basic form, it is formed by three stages, as shown in Figure 11.1: transmitter, communications channel, and receiver.

The transmission of the message can be carried out in baseband or bandwidth. The term baseband refers to the intrinsic frequency information of the signal, that is, the original frequency range it has of how it is generated in the information source (Lu, 1999).

Baseband transmission is characterized by transmitting the signal without shifting its frequency spectrum. It is used in applications that have a dedicated communication channel between source and destination, as in the case of the conventional telephone system, in which there is a single channel between the user and the telephone exchange. With the use of the dedicated channel, the signal is transmitted without the need to be translated into frequency, since the transmission of the information signal is performed from each user at a time.

Since the bandwidth of the communications channel is much larger than the bandwidth of a signals to be transmitted, the baseband transmission represents a signal propagation technique which uses the communications channel short of its communication capacity streaming.

In this scenario, bandwidth transmission enables the most efficient use of the communications channel. In it, the signals have their frequency spectra shifted to a region of higher frequency in the spectrum when comparing the frequency of the baseband signal.

The first step in a bandwidth communications system is performed by the transmitter, which has the function of adapting the signal so that it can be transmitted in a communication channel. In analog communications, the signal adaptation consists of the modulation process, which represents a

182 Continuous Wave Modulation

Figure 11.1 Basic block diagram of a communications system.

frequency offset of the signal to be transmitted in a communications channel, called the message signal.

The modulation provides several advantages to the transmission system, among which it is possible to mention the facility in Radio Frequency (RF) transmission. It is known that the parameters of frequency f, velocity v, and wavelength λ, are related by

$$v = f\lambda. \tag{11.1}$$

The modulation, upon taking off the frequency spectrum of the message signal, provides its transmission at high frequency. As the frequency increases, the wavelength decreases and, consequently the reduction of the antenna size required for signal transmission.

In such a scenario, for linear dipole, a signal can only be transmitted efficiently in a communications channel if the antenna size, L, is, at least, one-tenth of the wavelength, that is, (Balanis, 2016)

$$L = \frac{1}{10}\lambda. \tag{11.2}$$

Accordingly, it is necessary to increase the frequency by means of the modulation so as to achieve the reduction of the antenna size and the transmission of the signal in an appropriate way in a communications channel.

Another great advantage of the modulation process is the possibility of sending several signals simultaneously, from the technique called Frequency Division Multiplexing (FDM).

The FDM consists of a method of spectrum sharing which, by means of the modulation, the frequency spectrum of each message signal to be transmitted by the communications channel is shifted to a specific frequency range, so as to provide the transmission of several simultaneously without the overlap between them.

Figures 11.2(a) and 11.2(b) shows an example of FDM multiplexing and FDM demultiplexing, respectively. In this case, three signals have their offset baseband spectra, one for each transmission frequency (f_1, f_2 and f_3), defined from the specific application and bandwidths of the signals. In this

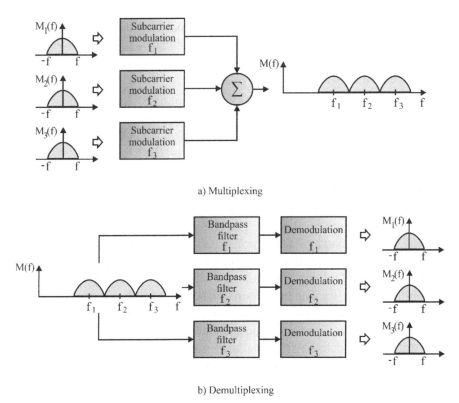

Figure 11.2 Simultaneous transmission of signals by time-division multiplexing: (a) frequency multiplexing; (b) frequency demultiplexing.

configuration, they can be transmitted simultaneously without overlapping of the spectra and, consequently, interference between them. At reception, the signals are obtained using bandpass filters, centered on the frequency of transmission and with cut-off frequency defined from the bandwidth of the transmitted signals (Forouzan, 2012).

Modulation also enables transmission bandwidth exchange for better performance against interference.

The literature has several different modulation techniques. It consists of the process of changing some parameters of a sinusoidal signal. It is usually considered a sinusoid whose angle is represented by an affine function, generically given by (Ding, 2010)

$$z(t) = A_z(t)\cos(\omega_z t + \theta_z(t)), \tag{11.3}$$

in which $A_z(t)$, ω_z and $\theta_z(t)$ represent respectively the amplitude, angular frequency, and phase of the generic sinusoidal signal. According to the parameters of the generic sinusoidal signal, the modulation can be realized in three distinct ways, they are (Jr, 2005):

1. Amplitude Modulation (AM): the amplitude $A_z(t)$ of the sinusoidal signal varies linearly with the message signal $m(t)$;
2. Frequency Modulation (FM): the frequency f_z of the sinusoidal signal varies linearly with the message signal $m(t)$;
3. Phase Modulation (PM): the phase $\theta_z(t)$ of the sinusoidal signal varies linearly with the message signal $m(t)$.

After the modulation process, the signal is capable of being transmitted by a communications channel. The communications channel has an analog nature and the message signal propagated by it suffers attenuation and distortion from sources of interference, besides being corrupted by random signals called noises. The communications channel can be classified as linear or non-linear, time-varying or time-invariant, and limited in range or power.

In its simplest form, the communications channel can be modeled by means of a noise type that is added to the message signal, has its probability distribution modeled by the Gaussian probability distribution function, and has its constant frequency spectral density for any spectrum frequency. This channel is called the Additive White Gaussian Noise (AWGN) channel.

In this scenario, the signal received at the receiver, $r(t)$, is given by the sum of the message signal transmitted, $s(t)$, with the noise $n(t)$, that is,

$$r(t) = s(t) + n(t). \qquad (11.4)$$

In analog communications, the receiver aims to reproduce the waveform of the transmitted signal as accurately as possible.

The amplitude, frequency and phase modulations are discussed below, with emphasis on their modulator and demodulator circuits.

11.1 Amplitude Modulation

In general, an analog modulation technique involves two signals. The first of these is the message signal, also known as the modulating signal, which represents the information to be transmitted by the communications channel. The second signal, known as the carrier signal or simply carrier, is the one that carries the information of the signal contained in the message signal through the communication channel.

11.1 Amplitude Modulation

Amplitude Modulation is characterized by the linear variation of the amplitude of the carrier signal according to the variation of the amplitude of the message signal.

Here, two types of amplitude modulation are analyzed: amplitude modulation with double sideband without carrier transmission (AM-DSB-SC) and amplitude modulation with double sideband and carrier transmission (AM-DSB).

11.1.1 Amplitude Modulation – Double Side Band-Supressed Carrier (AM-DSB-SC)

Consider a carrier signal, $c(t)$, is usually represented by a generic senoid given by

$$c(t) = A\cos(\omega_c t + \theta), \tag{11.5}$$

in which A consists of the amplitude of the signal, and ω_c and θ represent respectively the angular frequency and the phase of the carrier.

The amplitude modulation technique assigns the carrier the amplitude variation of the message signal by means of the multiplication of the signals, as shown in Figure 11.3.

Hence, the modulated signal, $s_{AM}(t)$, is given by

$$s_{AM}(t) = Am(t)\cos(\omega_c t), \tag{11.6}$$

in which $m(t)$ is the message signal with a bandwidth of B Hz. Here, the carrier phase is equal to zero to facilitate the exposure of the modulation, without affecting the results. The modulated signal presented in Formula 11.6 is in the form of double-sideband amplitude modulation and without carrier transmission, that is, AM-DSB-SC.

Figure 11.4(b) illustrates the signal modulated $s_{AM}(t)$ by a message signal with an arbitrary waveform shown in Figure 11.4(a). The modulated signal is obtained from the carrier within the envelope obtained by the

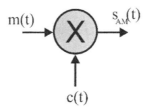

Figure 11.3 Scheme for AM-DSB-SC signal generation.

186 *Continuous Wave Modulation*

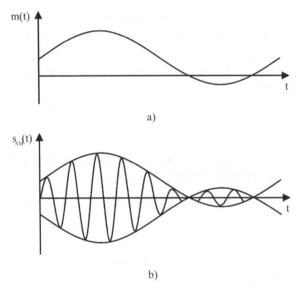

Figure 11.4 AM-DSB-SC modulation: (a) message signal; (b) modulated signal.

modulating signal and its mirrored signal. It is observed that the modulated signal undergoes a phase reversal whenever the message signal crosses the zero. Thus, the AM-DSB-SC signal envelope is different from the envelope of the message signal.

In communications systems, there are two limited resources under which any transmission techniques should consider: power and bandwidth (Ding, 2010).

The power P of a signal is obtained by means of

$$P = \lim_{T \to \infty} \frac{1}{T} \int_{-\frac{T}{2}}^{\frac{T}{2}} |x(t)|^2 dt. \tag{11.7}$$

Thus, from Formula 11.7, the modulated carrier power is given as follows

$$P = \lim_{T \to \infty} \frac{1}{T} \int_{-\frac{T}{2}}^{\frac{T}{2}} |Am(t)\cos(\omega_c t)|^2 dt. \tag{11.8}$$

Thus,

$$P = \lim_{T \to \infty} \frac{1}{T} \int_{-\frac{T}{2}}^{\frac{T}{2}} \left| \frac{1}{2}A^2 m^2(t) + \frac{1}{2}A^2 m^2(t)\cos(\omega_c t) \right| dt. \tag{11.9}$$

11.1 Amplitude Modulation

Since the message signal has a frequency much smaller than the second term of the integral, it is considered that the second term tends to zero when the period tends to infinity. Therefore, the power of the modulated carrier is given by

$$P_{AM} = \frac{A^2 P_M}{2}, \qquad (11.10)$$

in which P_M is the power of the message signal, given by (Leon-Garcia, 2008)

$$P_M = \frac{E[m^2]}{2}. \qquad (11.11)$$

The second parameter to be considered in the AM modulation technique is the bandwidth, which is obtained by analyzing the frequency characteristic of the modulated carrier.

By analyzing the frequency domain modulated signal, from the Fourier Transform of $s_{AM}(t)$,

$$S_{AM}(f) = \frac{A}{2}[M(f + f_c) + M(f - f_c)], \qquad (11.12)$$

in which $M(f)$ is the Fourier Transform of the message signal (Lathi, 2004).

Expression 11.12 suggests that the amplitude modulation process shifts the frequency spectrum of the message signal to the center frequency of the carrier, f_c, which is much larger than the baseband frequency of the message signal, that is,

$$f_c \gg B. \qquad (11.13)$$

The bandwidth required for transmission of the message signal in AM-DSB-SC can be obtained from the observation of Figure 11.5. Consider an arbitrary frequency spectrum of the message signal, shown in Figure 11.5(a), band-bounded at interval $-B \ll f \ll B$. The modulation process translates the spectrum of the message signal, from baseband to frequency $\pm f_c$, as shown in Figure 11.5(b). Thus, by observing the spectrum of the modulated

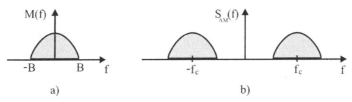

Figure 11.5 AM-DSB-SC modulation in the frequency domain: (a) message signal spectrum; b) modulated signal spectrum.

Continuous Wave Modulation

signal, it is realized that the bandwidth, B_T, necessary to transmit a message signal in AM-DSB-SC modulation is twice the bandwidth of the message signal, that is,

$$B_T = 2B. \qquad (11.14)$$

Synchronous or Coherent Detection

Demodulation is the process in which the message signal is obtained from the modulated signal. In the case of the AM-DSB-SC signal, it is performed from the synchronous demodulation, also known as coherent demodulation. This technique is thus termed as having a timing between the modulator and the demodulator.

This type of demodulation is used in modulation techniques in which the modulated signal has a phase inversion so that the envelope of the modulated signal does not fully correspond to the envelope of the message signal, as is the case with AM-DSB-SC modulation.

Figure 11.6 shows the block diagram of the synchronous demodulation. For signal demodulation, the modulated signal is initially multiplied by a locally generated signal which ideally must have the same phase and frequency characteristics of the sine wave signal using in the modulator as a carrier. Thereafter, the resulting signal is filtered by a low pass filter (Haykin, 2001).

In order to analyze coherent demodulation, consider that the receiver locally generates a sinusoidal signal with the same frequency as that generated in the transmitter, but with an arbitrary phase difference ϕ. Thus, the local oscillator must generate the sinusoidal signal, $c_1(t)$, given by (Haykin, 2001)

$$c_1(t) = A_1 \cos(\omega_c t + \phi), \qquad (11.15)$$

in which A_1 is the amplitude of the sine generated at the receiver.

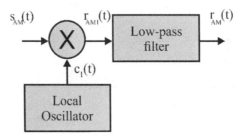

Figure 11.6 Block diagram of the synchronous demodulation.

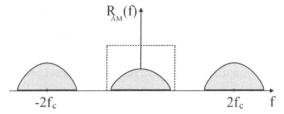

Figure 11.7 Frequency spectrum illustrating baseband component filtering.

The demodulation is done by multiplying the signal received by the receiver with the sine signal generated locally, that is,

$$r_{AM1}(t) = AA_1 m(t) \cos(\omega_c t) \cos(\omega_c t + \phi). \tag{11.16}$$

Using the trigonometric relation of the cosine multiplication,

$$r_{AM1}(t) = \frac{1}{2} AA_1 m(t) \cos(\phi) + \frac{1}{2} AA_1 \cos(2\omega_c t + \phi). \tag{11.17}$$

After the low pass filter,

$$r_{AM}(t) = \frac{1}{2} AA_1 m(t) \cos(\phi). \tag{11.18}$$

Figure 11.7 illustrates the resulting spectrum of the synchronous demodulation process, in which it can be seen that the message signal sent through the communications channel is obtained by filtering through the low pass filter, while the higher frequency focused spectra are filtered.

By analyzing the signal resulting from the filtering, it is observed that the detector provides an attenuated output of the message signal, by a factor equal to $cos(\phi)$. When the phase error ϕ is constant, $r_{AM}(t)$ is proportional to the message signal $m(t)$, and reaches its maximum and minimum values for a phase error of $\phi = 0$ and $\phi = \pm\pi/2$, respectively. When the phase error is $\phi = \pm\pi/2$ the value of $r_{AM}(t)$ is zero and represents the quadrature null effect.

Knowing then that the communication channel has analog nature, whose characteristics vary with time, the phase error must also vary randomly over time, resulting in undesirable effects on the demodulated signal. Thus, in order to have adequate synchronous demodulation, it is necessary for the receiver oscillator to locally generate a sine wave in phase and frequency synchronism with the carrier used to modulate the signal in the transmitter.

11.1.2 Amplitude Modulation – Double Side Band (AM-DSB)

The amplitude modulation with double sideband, or simply AM modulation, differs from the AM-DSB-SC modulation by sending through the communication channel the pure carrier, in addition to the modulated carrier.

The AM modulation is characterized by varying linearly and under a midpoint the amplitude of the carrier wave, $c(t)$. In addition to the modulated carrier, the pure carrier is also sent, resulting in the AM wave given by (Ding, 2010)

$$x_{AM}(t) = A[1 + \mu_{AM} m(t)] \cos(\omega_c t), \qquad (11.19)$$

in which μ_{AM} is a constant called amplitude sensitivity of the modulator that generates the signal modulated in amplitude. The AM wave can be divided into two terms: carrier, represented by the term $A \cos(\omega_c t)$, and lateral band, represented by the term $A\mu_{AM} \cos(\omega_c t)$.

Figure 11.8(b) illustrates the waveform of the amplitude-modulated carrier by a pure sine wave signal, $m(t) = \text{sen}(\omega_m t)$, shown in Figure 11.8(a). In this case, it was considered $\mu_{AM} = 0.5$.

In Figure 11.8(b) it can be seen that the envelope of $x_{AM}(t)$ has the same envelope as the baseband signal $m(t)$. However, in order to $x_{AM}(t)$ have the same message signal amplitude information, two conditions must necessarily be taken into account.

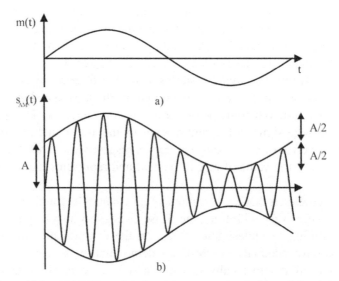

Figure 11.8 AM-DSB modulation: (a) message signal; (b) modulated signal.

The first condition is that the term $m(t)\mu_{AM}$ must always be less than unity, that is,
$$|\mu_{AM} m(t)| < 1, \tag{11.20}$$
for every instant of time t.

The transmission of the pure carrier along the modulated carrier has the objective of ensuring that the amplitude of the modulated wave is always positive and consequently the variation of the amplitude of the modulated wave has exactly the same amplitude variation of the message signal. Thus, the first condition ensures that the function $1 + m(t)\mu_{AM}$ is always positive and thus it is possible to express the AM wave envelope as $A[1 + m(t)\mu_{AM}]$.

An important observation is about the effect of the value of the amplitude sensitivity parameter, also known as the modulation index, $\mu_{AM}(t)$, in the modulated signal. In order for the first AM modulation condition to be satisfied, the modulation index must have a value in the range given by
$$0 \leq \mu_{AM} \leq 1, \tag{11.21}$$

If the modulation index was greater than the unit, give the name of overmodulation, and it results in the phase inversion of the carrier whenever the term $1 + m(t)\mu_{AM}$ crosses the zero. In the case of overmodulation, the AM wave exhibits a distortion in the envelope, and thus, does not exhibit the same amplitude variation of the message signal.

The modulation index represents the depth level that the message signal provides the carrier, and is defined by (Ding, 2010)
$$\mu_{AM} = \frac{m_{max} - m_{min}}{2A + m_{max} + m_{min}}, \tag{11.22}$$

in which m_{max} and m_{min} represent the maximum and minimum amplitudes of the message signal. If the modulation index is equal to unity, we have the situation of the total deepening of the message signal in the carrier, as shown in Figure 11.9.

The second condition in the AM modulation process to have a detectable envelope is that the carrier frequency f_c must be much larger than the largest frequency component of the message signal, B, that is,
$$f_c \gg B. \tag{11.23}$$

The effect of the AM modulation process on the frequency domain is obtained by the Fourier Transform of the modulated wave given by Formula

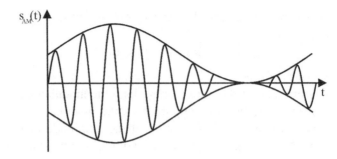

Figure 11.9 AM-DSB modulated signal with modulation index equal to unity.

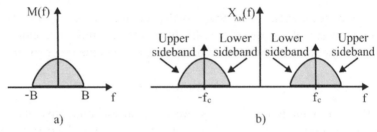

a) b)

Figure 11.10 Representation of frequency domain AM-DSB modulation: (a) spectrum of the baseband signal; (b) Spectrum of the AM-DSB modulated signal.

11.19. Thus, the spectrum of the amplitude modulated signal is given by

$$X_{AM}(f) = \frac{A}{2}[\delta(f+f_c) + \delta(f-f_c)] + \frac{A\mu_{AM}}{2}[M(f+f_c) + M(f-f_c)]. \quad (11.24)$$

Figure 11.10(a) illustrates the frequency spectrum of the message signal, limited in frequency in the range $-B \ll f \ll B$ and Figure 11.10(b) shows the spectrum $X_{AM}(f)$, in which it is possible to observe the frequency shift of the baseband spectrum to the center frequency of the carrier $\pm f_c$, and multiplied by the factor $A\mu_{AM}/2$, as well as the presence of impulses with factor-weighted $A/2$ also located at the center frequency of the carrier $\pm f_c$.

From the observation of the spectrum, two considerations can be made. The first of these is that the spectrum can be viewed from its sidebands. For the positive frequency side, the side of the spectrum above the carrier frequency is called the upper sideband, while the underside of the carrier frequency is called the lower sideband. Similarly, for negative frequencies, the side of the spectrum with frequency above the carrier frequency is called the lower sideband, and the side of the spectrum with frequency lower than the center frequency is called the upper sideband.

The second observation to be made on the AM wave spectrum is in relation to the bandwidth provided by the modulation process. It is noted that the negative frequency part of the baseband signal spectrum is shifted to the positive frequency part, resulting in the doubling of the bandwidth required to transmit the amplitude modulated signal. Thus, the bandwidth provided by the AM-DSB modulation, B_T, is given by

$$B_T = 2B. \tag{11.25}$$

In addition to the bandwidth provided by the AM modulation technique, it is necessary to analyze the power required for AM wave transmission through a communications channel. Since the AM signal can be separated into two terms, that is, carrier and sideband, the total power P_T required for the transmission is given by

$$P_T = \frac{A^2}{2} + \frac{P_M}{2}, \tag{11.26}$$

in which the first term of Formula 11.26 represents the power required to send the carrier, and the second term is the power required to send the sideband.

An important consideration of the power for AM wave transmission is that most of the total power generated at the transmitter is used to send the carrier. However, the message signal is present in the sideband.

Therefore, it is necessary to compute the power yield in the transmission of the signal modulated in amplitude. The yield consists of a power ratio, that is, between the power used to transmit the message, P_u, which is the power of the sideband, and the total power. Then the yield η is given by (Haykin, 2001)

$$\eta = \frac{P_u}{P_T}. \tag{11.27}$$

Asynchronous Demodulation and Comparison between AM-DSB-SC and AM-DSB

In this text, two configurations for AM modulation are presents, with carrier sending (AM-DSB) and without carrier sending (AM-DSB-SC). These two modulations must be analyzed in terms of demodulation, bandwidth, and power.

The principle of the AM-DSB is to send, in addition to the modulated carrier, the pure cover. This characteristic has the objective of providing a simpler demodulation technique, called asynchronous or envelope demodulation.

Envelope demodulation is an efficient way of detecting the message signal and implemented more simply than the synchronous modulation, from a nonlinear device and a low-pass filter. The pure carrier sent represents a DC level in the carrier carrying the information, making its amplitude variation oscillate only on the positive part around an average value, given exactly by the amplitude of the carrier, that is, by the DC level. Thus, as mentioned, the envelope of the modulated wave corresponds exactly to the envelope of the message signal, being possible to detect it with a simple circuit, which is the envelope detector.

In general, both of the modulation techniques can be demodulated synchronously. This choice is also related to the Signal-to-Noise Ratio (SNR) of the transmission. When the SNR is low, it is convenient to use the synchronous demodulation. However, since power is used in the transmitter to send the pure carrier, it is convenient to make the choice for simpler demodulation.

Regarding power, as mentioned, much of the power destined for AM-DSB modulation is intended for pure carrier transmission, which has no information and is only used for demodulation purposes. In the case of AM-DSB-SC modulation, all the power generated in the transmitter is destined to transmit the message signal information. However, its demodulation is performed from more complex circuits, which is the synchronous demodulation.

The two amplitude modulation settings require twice the bandwidth of the message signal for sending the information over the communication channel. In the frequency spectrum given by the modulations, observed that there is asymmetry with respect to the central frequency. Thus, by the magnitude and phase information of one sideband, it is possible to determine the other sideband, meaning that in practice it is necessary to send only one of the sidebands and the communication channel must provide the same bandwidth of the message signal. Thus, can be said that the amplitude modulations AM-DSB-SC and AM-DSB waste frequency.

11.2 AM Modulators Circuits

Among all available modulation techniques, amplitude modulation stands out for its simplicity of implementation. The following are the main circuits used in the amplitude modulation and demodulation process.

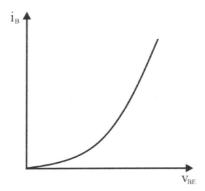

Figure 11.11 Voltage x current curve characteristic of the diode or common emitter transistor.

11.2.1 Quadratic Modulator

To obtain the signal modulated in AM-DSB, the quadratic modulator has its principle based on the characteristic of a non-linear element, which can be a common-emitter transistor or diode, in which the voltage versus current curve presents exponential behavior, given by Expression 11.28 and illustrated in Figure 11.11 (Gomes, 1985).

It is known that an exponential function can be decomposed by the Taylor series, that is,

$$e^x = 1 + x + \frac{x^2}{2} + \frac{x^3}{6} + \cdots \frac{x^n}{n!}. \tag{11.28}$$

However, if the transistor is polarized in order to restrict its operation in the non-linear region of the current-voltage curve, the multiplication between the carrier and the modulating signal occurs in the region of the curve that can be approximated by a parabola. In this case, for use in the quadratic modulator, the Expression 11.28 can be reduced to

$$e^x = 1 + x + \frac{x^2}{2}. \tag{11.29}$$

Since the transistor provides the following voltage and current characteristics

$$i_c = \beta i_B. \tag{11.30}$$
$$i_B = f(v_{BE}) \tag{11.31}$$

in which i_c is the current at the collector, i_b is the current at the base in the transistor, v_{BE} is the voltage between the base and emitter of the transistor

196   Continuous Wave Modulation

and β is a dimensionless known as gain of current. So, we can rewrite Formula 11.30 as follows

$$i_c = a_1 + a_2 v_{BE} + a_3 v_{BE}^2, \quad (11.32)$$

in which a_1, a_2 and a_3 are constant and the voltage between the base and the emitter is given the sum of the carrier signal, $c(t)$, and by the modulating signal, $m(t)$, that is

$$v_{BE} = c(t) + m(t). \quad (11.33)$$

with

$$c(t) = A\cos(\omega_c t). \quad (11.34)$$

Since the transistor is a non-linear device, the voltage v_{BE} generates a collector current given by

$$i_c(t) = a_1 + a_2(c(t) + m(t)) + a_3(c(t) + m(t))^2. \quad (11.35)$$

After the mathematical manipulations, the current in the collector is given by

$$i_c(t) = a_1 + a_2 A\cos(\omega_c t) + a_2 m(t) + \frac{a_3 A^2}{2} + \frac{a_3 A^2}{2}\cos(2\omega_c t)$$
$$+ 2a_3 A m(t)\cos(\omega_c t) + m(t)^2. \quad (11.36)$$

To generate the modulated signal at the carrier frequency, the signal given by Expression 11.36 must be input to a center band filter centered in ω_c, resulting in

$$i_c(t) = a_2 A\cos(\omega_c t) + 2a_3 A m(t)\cos(\omega_c t). \quad (11.37)$$

Figure 11.12 illustrates the power spectrum of the current $i_c(t)$, after passage through the bandpass filter.

Figure 11.13 illustrates the circuit corresponding to the quadratic modulator using the transistor as a nonlinear element. In this circuit, the signals

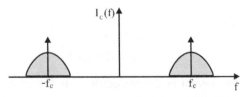

Figure 11.12   Frequency spectrum of $i_c(t)$.

11.2 AM Modulators Circuits

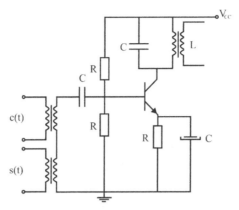

Figure 11.13 Electrical circuit of the quadratic modulator with transistor.

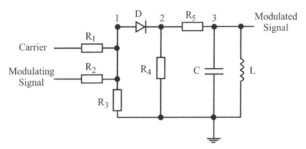

Figure 11.14 Electrical circuit of the quadratic modulator with diode.

are summed and represent the transistor base-emitter voltage. The collector current is then the input of the bandpass filter, represented by the LC circuit, for the generation of the AM-DSB modulated signal.

Another possible configuration of the quadratic modulator is illustrated in Figure 11.14 with the use of the diode. The circuit can be observed in three parts. At point 1, the message signal is added to the carrier, from the use of resistors, and then passes through a non-linear device, in this case, the diode. At point 2, the signal has the quadratic form, which is passed through a bandpass filter (resonant circuit LC) at point 3, for the generation of the AM-DSB modulated signal.

11.2.2 Modulator by Switching or Synchronous

The synchronous is based on the principle that a signal being sampled by a periodic signal results in a series of harmonics that can be conveniently selected by a bandpass filter.

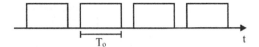

Figure 11.15 Periodic pulse train.

Consider periodic pulse train, as shown in Figure 11.15, which in the modulation process represents the carrier signal, has timing at each $T_o = 1/F_o$ and can be mathematically represented by its Fourier series as

$$c(t) = c_0 + c_1 \cos(\omega_c t) + c_2 \cos(3\omega_c t) + c_3 \cos(5\omega_c t) + \cdots \quad (11.38)$$

in which $c_n, n = 1, 2 \ldots \infty$ are the coefficients of the Fourier series (Lathi, 2004).

Performing the AM-DSB modulation process,

$$s_{AM}(t) = s(t) + c(t) = m(t)c(t) + c(t). \quad (11.39)$$

In the frequency domain,

$$\begin{aligned} S_{AM}(f) = & \left[c_0 M(f) + \frac{c_1}{2}[M(f + f_c) + M(f - f_c)] \right. \\ & \left. + \frac{c_2}{2}[M(f + 3f_c) + M(f - 3f_c)] \right] \cdots \\ & + c_0 + \frac{c_1}{2}[\delta(f + f_c) + \delta(f - f_c)] \\ & + \frac{c_2}{2}[\delta(f + 3f_c) + \delta(f - 3f_c)] \cdots \end{aligned} \quad (11.40)$$

The Formula 11.40 presents the signal spectrum resulting from the modulator by switching, in which it is possible to observe that the actuation of the bandpass filter selecting the signal of interest. After the filter, the resulting signal is given by

$$S_{AM}(f) = \frac{c_1}{2}[\delta(f + f_c) + \delta(f - f_c)] + \frac{c_1}{2}[M(f + f_c) + M(f - f_c)]. \quad (11.41)$$

The electrical circuit that implements the modulator by switching is shown in Figure 11.16. In the circuit, the resistors R_0, R_1 e R_2 configure a resistive adder, while the diode assumes the key role, which opens and closes periodically at a frequency f_o. The capacitors and the inductor form the bandpass filter, or tuned circuit, centered on the frequency of forming interest of the AM-DSB modulated signal.

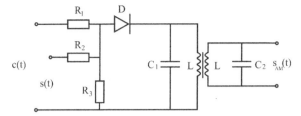

Figure 11.16 Electrical circuit of the modulator by switching with diode.

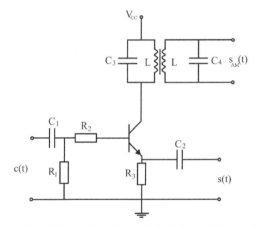

Figure 11.17 Electrical circuit of the modulator by switching with transistor.

Like the quadratic modulator, the synchronous modulator can also be implemented using the transistor as a synchronous switch, as shown in Figure 11.17. In this case, the effect of the synchronous switch is obtained when the transistor oscillates between its conducting and cutting state and so it cannot be biased with a DC level.

11.2.3 Balanced Modulator

The balanced modulator is used to generate amplitude modulated signals without carrier transmission (AM-DSB-SC). Figure 11.18 illustrates the electronic circuit of the balanced modulator, in which it is possible to observe that it is formed by the junction of two quadratic modulators and a bandpass filter (Gomes, 1985).

As in the quadratic modulator, the balanced modulator uses nonlinear elements (in this case, two transistors), which must be operating in the region of the curve that is approximately non-linear.

200   Continuous Wave Modulation

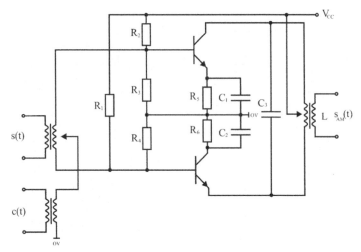

Figure 11.18   Electrical circuit of the balanced modulator with transistor.

The signal of interest of each transistor is the current signal of the collector. For the transistor T_1, the collector current is given by

$$i_{c1}(t) = d_1 + d_2[c(t) + m(t)] + d_3[c(t) + m(t)]^2. \tag{11.42}$$

One feature of the balanced modulator is that the capacitors C_1 and C_2 have negligible capacitances for alternating signals. Thus, the voltage between the base and the emitter of the collector can also be represented by the difference of the modulated carrier and pure carrier. Thus, for the second transistor T_2, the collector current is given by

$$i_{c2}(t) = d_1 + d_2[c(t) - m(t)] + d_3[c(t) - m(t)]^2. \tag{11.43}$$

The current, i, which reaches the inductor L is given by

$$i = i_{c1} - i_{c2}. \tag{11.44}$$

By making the substitution,

$$i(t) = 2d_2 m(t) + 2d_3 m(t) \cos(\omega_c t). \tag{11.45}$$

After passing the current through a bandpass filter, represented by the LC circuit, the modulated signal AM-DSB-SC, given by Formula 11.46 and its frequency spectrum is illustrated in Figure 11.19.

$$i(t) = 2d_3 m(t) \cos(\omega_c t). \tag{11.46}$$

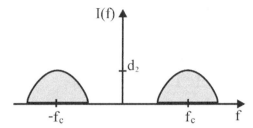

Figure 11.19 Electrical circuit of the balanced modulator with transistor.

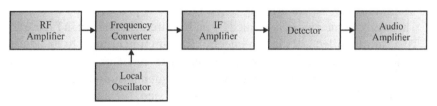

Figure 11.20 Block diagram of the superheterodyne receptor.

11.3 AM Demodulator

The demodulation of a signal consists of extracting the message signal from the signal received at the receiver of a communications system. There are two types of demodulation: asynchronous or by envelope and the synchronous. The choice of demodulation type should be based on the type of amplitude modulation used to transmit the signal and the Signal-to-Noise Ratio (SNR).

The widely used receptor in the literature and used to demodulate amplitude-modulated signals is the superheterodyne receptor, the block diagram of which is shown in Figure 11.20.

The superheterodyne receiver was designed with the purpose of providing signal reception in applications involving different transmitting stations. From the frequency translation of the carrier from the transmitting station to an intermediate frequency, it is able to select and demodulate signals being transmitted in different carriers.

This receiver can be used to receive AM and FM signals. Your project includes steps of frequency synchronization of the carrier wave, filtering, and amplification. One of the great advantages of using the superheterodyne receiver is that because it has the step of converting the carrier frequency to an intermediate frequency, smaller than the original frequency of signal transmission, it does not require the use of a highly selective and variable tunable filter, which are circuits with greater complexity.

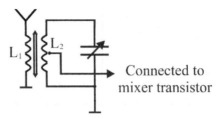

Figure 11.21 Electrical circuit of the tuned filter.

Initially, the signal arriving at the receiver is picked up by a tuned filter, whose electronic circuit is formed by an LC circuit, as shown in Figure 11.21. This step is called a Radio Frequency (RF) amplifier, tuned to an adjustable frequency f_c from the capacitor value. The inductor, in this case, may represent the receiving antenna itself or the antenna coupling.

After the signal is picked up by the RF step, the signal is input from a mixer circuit, which has the function of converting the signal to an intermediate frequency, which is the demodulation frequency used by the detector to convert the signal to the band frequency base. Mixing circuits have already been discussed in Chapter 9, as well as some of the possible types of oscillators. Figure 11.22 illustrates the Colpitts oscillator, representing the local oscillator, next to the mixer circuit (Gomes, 1985) (Lee, 1998).

The signal is then amplified by the IF amplifier and then converted to baseband by detector circuits. Because they may have low amplitude, the baseband signals are amplified.

The following are the most common types of detection: envelope demodulation, quadratic detector, and synchronous detector.

11.3.1 Envelope Demodulation

Envelope detection is the simplest among amplitude modulated signal demodulation processes. It should only be used when, in addition to the message signal, the pure carrier signal is also transmitted, in order to make the envelope of the modulated carrier always positive, from the inclusion of a DC level.

This type of detection is performed by passing the modulated signal through a non-linear device followed by a low pass filter to eliminate high frequencies. Figure 11.23 shows the electrical circuit of the envelope detector, which uses the diode as a non-linear device (Campos, 2015).

The working principle of this circuit is simple. The diode acts as a synchronous key and the RC circuit as a low-pass filter.

11.3 AM Demodulator

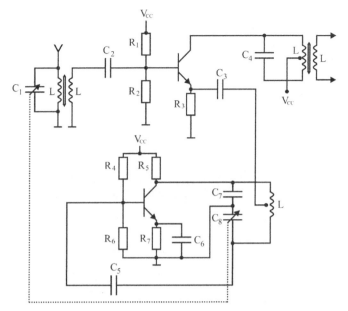

Figure 11.22 Colpitts oscillator with the mixer circuit.

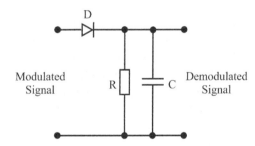

Figure 11.23 Electrical circuit of the envelope detector.

The signal arriving at the detector $s'_{AM}(t)$, passes through the diode, which together with the resistor rectifies the modulated signal, eliminating its negative parts, as shown in Figure 11.24(a).

By including the capacitor parallel to the resistor, the RC circuit representing the low-pass filter, provides the signal envelope as shown in Figure 11.24(b), that functions as a circuit in which the capacitor charges and discharges along the signal envelope. That is, the capacitor is charged with the peak value of the pulses that pass through the diode and is discharged through the resistor when the pulse voltage drops to zero.

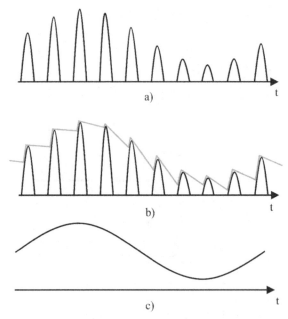

Figure 11.24 Envelope detector action: (a) Signal after diode; (b) Signal with low pass filter action; (c) Demodulated signal.

It is important to emphasize that the choice of the resistance and capacitance values must be made in order to make the RC constant suitable for the extraction of the envelope. The RC constant is related to the discharge of the capacitor. If it is very large, that is, greater than the maximum variation of the modulated signal $1/W$, the capacitor discharge is very slow and the output of the circuit has a very high DC ripple component.

On the other hand, if the RC constant is very small, that is, less than a carrier period, $1/f_c$, the capacitor discharge happens very fast, the output of the circuit will have a waveform similar to the shape of the modulated carrier, which is also not of interest to the detection process. Thus, it is important that the time constant satisfies the inequality

$$\frac{1}{f_c} \ll RC \ll \frac{1}{W}. \tag{11.47}$$

11.3.2 Quadratic Detector

Like the quadratic modulator, the quadratic demodulator uses a nonlinear element to demodulate the signal. The basic difference between the circuits is that the demodulator makes use of a low pass filter, while the modulator filters the signal at high frequencies with the bandpass filter.

Admit that the signal arriving at the detector, $r(t)$, is given by

$$r(t) = A_{r1} \cos(\omega_c t) + A_r m(t) \cos(\omega_c t), \tag{11.48}$$

in which A_{r1} and A_r are, respectively, the receiving amplitude carrier and the modulated signal.

The nonlinear element must be configured to work in an approximately quadratic region, given by

$$e(t) = k_0 + k_1 e(t) + k_2 e^2(t), \tag{11.49}$$

in which k_0, k_1 and k_2 are coefficients of the quadratic function. Passing $r(t)$ through the quadratic device, we have $r'(t)$ given by

$$r'(t) = k_o + k_1(A_{r1} \cos(\omega_c t) + A_r m(t) \cos(\omega_c t))$$
$$+ k_2(A_{r1} \cos(\omega_c t) + A_r m(t) \cos(\omega_c t))^2. \tag{11.50}$$

After the mathematical manipulations and the passage through a low pass filter, centered on the frequency of the message signal, we have

$$r''(t) = k_o + \frac{A_{r1}^2}{2} + \frac{A_r A_{r1}}{2} m(t), \tag{11.51}$$

in which the two first terms in Expression 11.51 represent a level DC and the third represents the message signal.

11.3.3 Synchronous Detector

The principle of operation of the synchronous detector is similar to the synchronous modulator, with the difference that the detector uses a low pass filter, while the modulator uses a bandpass filter.

The synchronous detector uses a non-linear element as a synchronous key, which may be a diode or transistor. Thus, the signal $r(t)$, received in the detector, is sampled by the synchronous key resulting in the signal $r'_1(t)$ given by

$$r'_1(t) = r(t) \times c(t), \tag{11.52}$$

in which $c(t)$ is the series of the train of pulses that represents the synchronous key. In this way,

$$r'_1(t) = (c_0 + c_1 \cos(\omega_c t) + c_2 \cos(3\omega_c t) + c_3 \cos(5\omega_c t) + \cdots)$$
$$\times (A_{r1} \cos(\omega_c t) + A_r m(t) \cos(\omega_c t). \tag{11.53}$$

After the mathematical manipulations and the passage through the low pass filter centered on the frequency of the message signal, we have

$$r''_1(t) = \frac{c_1 A_{r1}}{2} + \frac{A_r c_1}{2} m(t). \tag{11.54}$$

in which the two first terms in Expression 11.54 represent a level DC and the third represents the message signal.

11.4 Angular Modulation

After AM modulation appeared, the researchers sought to find a modulation that would enable the transmission of the message signal with a lower bandwidth than that required by AM modulation.

The reason for developing a new modulation was based on the noise characteristic being proportional to the bandwidth of the modulated signal. Thus, a modulation that would provide smaller bandwidth than that provided by AM modulation would be more robust against the noise of communications.

Angular modulation emerged as an alternative to amplitude modulation with a goal of providing a bandwidth less than the bandwidth provided by amplitude modulation. However, what has been achieved is a modulation that requires at least twice the bandwidth of the message signal and that in practice the bandwidth is much larger than that required to transmit the message signal with the AM modulation. Despite this higher bandwidth characteristic, angular modulation has important advantages compared to AM modulation, such as being able to perform bandwidth adjustment and provide immunity to non-linear devices.

As seen, the information carrier wave has two parameters in which it is possible to vary for the transmission of the message signal. The AM modulation has the characteristic of varying the amplitude of the carrier, while the angular modulation inserts the information to be transmitted at the angle of the carrier, that is, the angle of the carrier wave is varied according to the message signal.

Consider initially a generic sinusoidal carrier, $s_A(t)$, given by

$$s_A(t) = A_e \cos[\theta_i(t)], \tag{11.55}$$

in which A_e and $\theta_i(t)$ are respectively the amplitude and angle of the carrier, represented by a generic function.

11.4 Angular Modulation

Now consider a carrier $s'_A(t)$ with an angle given by a linear function, that is

$$s'_A(t) = A_e \cos[\omega_c t + \phi_o], \tag{11.56}$$

in which ω_c and ϕ_o are respectively the angular frequency and the initial phase of the carrier.

In some time interval from t a $t + \Delta t$, the angle of the carrier $s'_A(t)$ tangentiates the angle of $s_A(t)$ and, if $\Delta t \to 0$ the two carriers are equal. In this time interval, the slope of the line $\omega_c t + \phi_o$ is given by the angular frequency ω_c, that is, for a given $\Delta t \to 0$ a the instantaneous angular frequency is given by ω_c. Generally considering any interval of time that tends to zero, the instantaneous frequency f_i is given by

$$f_i = \frac{1}{2\pi} \frac{d\theta_i(t)}{dt}. \tag{11.57}$$

The angular modulation can be configured with its angle varying from infinite relationships with the message signal. Consider, however, a generalized angular modulation given by

$$s_{em}(t) = A_e \cos(\omega_c t + \psi(t)), \tag{11.58}$$

in which $\psi(t)$ is a function proportional to the message signal. Assume that $\psi(t)$ is obtained by a linear and time-invariant filter, with impulse response $h(t)$. Thus, the angle modulated generic signal is given by

$$s_{em}(t) = A_e \cos\left(\omega_c t + \int_{-\infty}^{t} m(\alpha) h(t-\alpha) d\alpha\right). \tag{11.59}$$

Although there are infinite possibilities of angular modulation, two types are usually considered: phase modulation and frequency modulation.

Phase Modulation (PM): is the shape of the angular modulation at which the angle $\theta_i(t)$ is varied linearly with the message signal. In this case, $h(t) = k_p \delta(t)$, and for the PM modulation,

$$\theta_i(t) = 2\pi f_c t + k_p m(t), \tag{11.60}$$

in which k_p is the phase sensitivity of the modulator, expressed in radian per volt, considering that the message signal is a voltage source.

The phase-modulated carrier $s_{PM}(t)$ is given in the time domain by

$$s_{PM}(t) = A \cos[2\pi f_c t + k_p m(t)]. \tag{11.61}$$

208 Continuous Wave Modulation

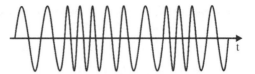

Figure 11.25 Frequency modulated carrier.

Frequency Modulation (FM): is the shape of the angular modulation in which the angle $\theta_i(t)$ is varied linearly with the integral of the message signal. In this case, $h(t) = k_f u(t)$, and for the PM modulation,

$$\theta_i(t) = 2\pi f_c t + k_f \int_0^t m(\tau)d\tau, \qquad (11.62)$$

in which k_f is the frequency sensitivity of the modulator, expressed in hertz per volt, assuming that the message signal is a source of voltage.

The frequency modulated carrier $s_{FM}(t)$ is given in the time domain by

$$s_{FM}(t) = A\cos\left[2\pi f_c t + 2\pi k_f \int_0^t m(\tau)d\tau\right]. \qquad (11.63)$$

Figure 11.25 illustrates the waveform of the FM modulated carrier for a generic message signal.

In addition to the instantaneous frequency, the angular modulation has two more important parameters: maximum phase deviation and maximum frequency deviation.

The maximum phase deviation $\Delta\theta$ is given by

$$\Delta\theta = 2\pi k_p \max\{|\phi(t)|\}. \qquad (11.64)$$

The frequency deviation Δf consists of the maximum distance of the instantaneous frequency of an FM wave from the carrier frequency and is given by

$$\Delta f = k_f \max\{|m(t)|\}. \qquad (11.65)$$

A fundamental characteristic of an FM wave is that the frequency deviation is proportional to the amplitude of the modulating wave and is independent of the modulation frequency, or of the carrier.

The study of a modulation scheme involves observing the two limited resources in a communications system: bandwidth and power.

Next, the bandwidth for the FM signal is analyzed. Since the signals in FM and PM are inseparable, the entire development shown below is also useful for PM modulation.

11.4.1 Narrow-Band Angle Modulator

Angular modulation can also be called exponential modulation since by its nature it can be represented exponentially by the Euler relation. Thus, the angular modulation can be represented by

$$s'_{FM}(t) = Ae^{j[2\pi f_c t + 2\pi k_f \int_0^t m(\tau)d\tau]}. \tag{11.66}$$

Admit that

$$z(t) = 2\pi k_f \int_0^t m(\tau)d\tau. \tag{11.67}$$

Expanding the exponential series of the FM signal,

$$s'_{FM}(t) = A[1 + jk_f z(t) - \frac{k_f^2}{2!}z^2(t) + \frac{k_f^3}{2!}z^2(t) + \cdots + j^n\frac{k_f^n}{n!}z^n(t)]e^{j\omega_c t}. \tag{11.68}$$

Knowing that the FM wave consists of the real part of $s'_{FM}(t)$,

$$s_{FM}(t) = A\left[\cos(\omega_c t) - k_f z(t)\operatorname{sen}(\omega_c t) - \frac{k_f^2}{2!}z^2(t)\cos(\omega_c t)\right.$$
$$\left. + \frac{k_f^3}{3!}z^3(t)\operatorname{sen}(\omega_c t) + \cdots\right]. \tag{11.69}$$

The signal $s_{FM}(t)$ is formed by a pure carrier and terms modulated by the signal $z(t)$. Tha modulated terms provide infinite bandwidth, since the spectrum provided by the term proportional to $z^n(t)$ is limited to nB.

To circumvent the infinite band characteristic of FM modulation, the terms that provide the part of the higher power spectrum are considered, and thus is obtained the modulation with limited bandwidth.

Narrowband FM is characterized by having a very small frequency deviation, that is,

$$|k_f z(t)| \ll 1. \tag{11.70}$$

With this condition, the FM signal becomes a linear signal, given by

$$s_{FM}(t) \approx A[\cos(\omega_c t) - k_f z(t)\operatorname{sen}(\omega_c t)]. \tag{11.71}$$

Similarly, the signal in PM is given by

$$s_{PM}(t) \approx A[\cos(\omega_c t) - k_p m(t)\operatorname{sen}(\omega_c t)]. \tag{11.72}$$

210 Continuous Wave Modulation

From the narrowband FM and PM signals, it is possible to observe similarity with the carrier-amplitude amplitude modulation (AM-DSB). Note that both modulations have the pure carrier term and a carrier modulated term by a measure of the message signal. Thus, like the AM-DSB modulation, narrow-band angular modulation has bandwidth given by twice the maximum message signal frequency component. The difference between the modulations is a 90° phase difference of the modulated carrier, as well as the carrier wave parameter used to allocate the message signal.

11.4.2 Wide-Band Angle Modulator

In practice, because it has a very small frequency deviation, the narrow-band FM cannot be used since the frequency deviation does not allow sufficient frequency variation to allocate the message signal.

To circumvent this situation, one has the wide-band FM, which, as is possible see below, provides a much wider bandwidth than the need for amplitude modulation. The bandwidth of an FM signal can be obtained by observing the sampled version of the message signal.

Admit the message signal $m(t)$ to be sampled by rectangular pulses, as shown in Figure 11.26(a). The sampling must be performed according to the sampling theorem, which states that the frequency components of the signal are conserved when the sampling rate is, at least, twice the maximum frequency of the continuous signal. Thus, if the largest frequency component of the message signal is given by B Hz, the sampling rate is given by $2B$ samples / s and the sampling period is given by $1/2B$.

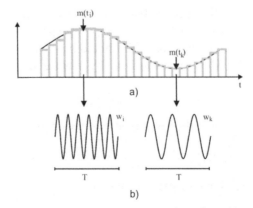

Figure 11.26 (a) message signal partition in rectangular pulses; (b) representation of each rectangular pulse by an FM signal.

11.4 Angular Modulation 211

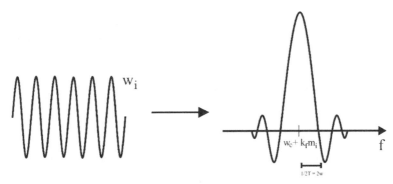

Figure 11.27 Fourier transform of the FM wave corresponding to each rectangular pulse.

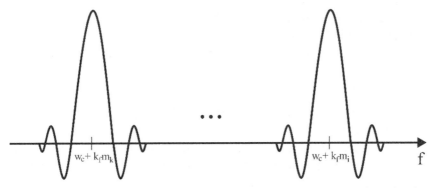

Figure 11.28 Illustration of the sync functions generated by each rectangular pulse, from the lowest instantaneous frequency to the highest instantaneous frequency.

For each rectangular pulse, there is an FM wave associated with it, with a given instantaneous frequency. Thus, the ith rectangular pulse is represented by an FM wave with the ith instantaneous frequency, as shown in Figure 11.26(b).

The interpretation in the frequency domain of the FM carrier is obtained by considering the junction of the spectrum of each FM wave referring to each rectangular pulse.

According to the Fourier Transform, the frequency domain representation of a rectangular pulse is given by the sync, as is illustrated in Figure 11.27. Figure 11.28 illustrates the spectral occupancy of the FM wave, which can be obtained from the sum of the sync functions, each situated at its instantaneous frequency obtained according to the message signal (Ding, 2010).

The bandwidth required for transmission of the message signal by means of the FM carrier is obtained by subtracting the largest frequency component by the smallest frequency component, that is

$$B_{FM} = (f_c + \Delta f + 2B) - (f_c - \Delta f - 2B), \quad (11.73)$$

or,

$$B_{FM} = 2\Delta f + 4B. \quad (11.74)$$

Thus, the bandwidth provided by the angular modulation is given by

$$B_{FM} = 2(\Delta f + 2B). \quad (11.75)$$

Formula 11.75 is known as the Carson Rule. Note that if the condition is assumed for a narrow band FM, the bandwidth supplied is twice that provided by narrowband FM (or AM modulation). Thus, an adjustment was made to the Carson rule, so that the bandwidth of the angular modulation is given by

$$B_{FM} = 2(\Delta f + B). \quad (11.76)$$

in which, the frequency deviation for the FM is $k_f \max\{|m(t)|\}/2\pi$ and for the PM is $k_p \max\{|m'(t)|\}/2\pi$.

An important parameter in angular modulation is the modulation index, which provides a relation between how much the angular modulation is varying the carrier frequency, with the baseband frequency, B of the carrier. The modulation index β is given by (Alencar and Rocha Jr, 2020)

$$\beta = \frac{\Delta f}{B}. \quad (11.77)$$

The Carson rule can, then, be expressed with the modulation index, as

$$B_{FM} = 2B(\beta + 1). \quad (11.78)$$

Power:

Since the amplitude of a phase-modulated wave or frequency is constant, that is, equal to the amplitude of the carrier, for every instant of time and independent of the sensitivity factors, the power P of the wave with angular modulation is given by

$$P = \frac{A^2}{2}. \quad (11.79)$$

11.5 FM Modulator Circuits

The generation of the signal in FM and PM can be done in two ways: indirect generation and direct generation.

11.5.1 FM Wave Indirect Generation

The indirect generation was developed by Armstrong and initially consists of the generation of narrowband FM too, then, generate the wide-band FM.

As shown in Section 11.4.1, narrowband FM and PM have a format very similar to AM-DSB, with the exception of phase difference between modulated carriers. Considering the relationship between the two modulation, it is possible to obtain the narrow band FM and PM from the AM-DSB-SC modulator, with the inclusion of a local oscillator to generate the carrier and a phase shift.

Figure 11.29 shows the block diagram for the narrowband PM generation. In it, the signal $m(t)$ is input from an AM-DSB-SC modulator, which has at its output the modulated carrier. It is observed that there is a local oscillator, generating a high-frequency carrier, which undergoes a phase separation of $90°$ and is input to the modulator circuit.

Figure 11.30 illustrates the FM wave generation. The narrowband FM generation scheme follows the same narrow band PM principle, with the difference that the measurement of the message signal is obtained with the inclusion of an integrator circuit.

After generating the narrow band FM or PM, it is possible to generate the wideband FM or PM from the use of frequency multiplier circuits and

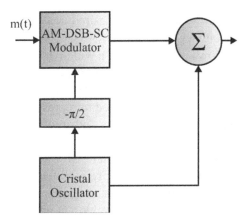

Figure 11.29 Block diagram for narrowband PM generation.

214 Continuous Wave Modulation

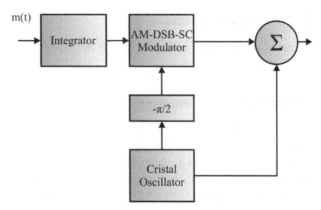

Figure 11.30 Block diagram for narrowband PM generation.

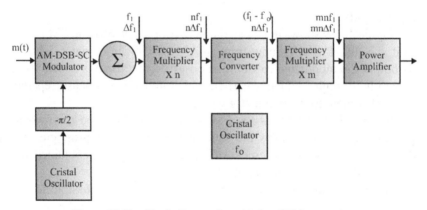

Figure 11.31 Block diagram for wide-band PM generation.

frequency converters, to achieve the desired center frequency of the carrier and the desired frequency deviation.

Figure 11.31 shows an example of a wide-band PM signal generation. In it, a narrow band PM wave with generic central frequency f_1 is generated initially and a generic standard deviation given by Δf_1. The center frequency and frequency offset need to be adjusted to represent a wide-band PM wave. For this, a suitable sequence of mixers and frequency converters should be included in the design.

After generation of the narrowband PM wave, a frequency multiplier circuit can be used, which will multiply both the center frequency and the frequency deviation by a factor n, resulting in a new center frequency and frequency deviation given, respectively, by nf_1 and $n\Delta f_1$.

Thereafter, a frequency converter, which is used of a local oscillator with frequency f_o, can be used, and has the characteristic of converting the center frequency to another frequency, without changing the frequency deviation. Now, the new frequency is given by $n(f_1 - f_o)$. It is still possible to use a second frequency multiplier to obtain the values $mn(f_1 - f_o)$ for the center frequency and $mn\Delta f_1$ for frequency deviation, thus obtain a wide-band PM wave.

11.5.2 FM Wave Direct Generation

Direct generation is performed using the voltage controlled oscillator (VCO), with the message signal $m(t)$ used with an external voltage source that controls the oscillator frequency variation.

The instantaneous frequency of the oscillator $f_i(t)$ is then given by

$$f_i = f_c + k_f m(t). \tag{11.80}$$

The VCO can be implemented by LC-type circuits (Hartley or Colpitts) or circuits with operational amplifiers. In the case of LC circuits, the frequency variation is obtained from the variation of the capacitance according to the following expression

$$C = C_0 + bm(t), \tag{11.81}$$

in which b is an adjustment parameter of the capacitance variation.

As mentioned in Chapter 9, the resonant frequency of an LC oscillator is given by

$$f_o = \frac{1}{\sqrt{RC}}. \tag{11.82}$$

The relation between the resonant frequency and the modulating signal is obtained by

$$f_o \approx \frac{1}{\sqrt{RC}}\left[1 + \frac{bm(t)}{2C_0}\right], \tag{11.83}$$

with $\frac{bm(t)}{2C_0} \ll 1$.

The maximum capacitance deviation ΔC is given by

$$\Delta C = C - C_0 = bm(t). \tag{11.84}$$

Therefore, is observed that the variation in the capacitance results in the variation of frequency and, consequently, in the generation of the FM signal.

11.6 FM Demodulator Circuits

The FM demodulation can be performed by means of a frequency discriminator or by PLL circuits.

11.6.1 FM Demodulation with PLL

The PLL circuit is widely used in communications circuits and one of the applications is in the demodulation of frequency-modulated signals. Currently, PLL is considered the best form of FM demodulation due to linearity, good signal-to-noise ratio, and frequency selectivity.

The use of PLL in FM signal demodulation stems from the fact that it is able to rapidly track the frequency changes of an input signal, in this case, the frequency-modulated carrier.

In order for demodulation to occur by PLL, the frequency shift of the modulated carrier must be within the capture range.

The operation of the PLL circuit has already been discussed in Chapter 11. Figure 11.32 illustrates the PLL in the FM demodulation configuration. Note that, unlike the PLL shown in Chapter 11, in which the output of the PLL is the output of the VCO, the output of the PLL as FM demodulator is the output of the mesh filter.

Initially, the VCO output signal frequency should be set to be the same as the center frequency of the FM carrier. Since the frequency of the carrier varies around the center frequency and according to the modulating signal, that is, the message signal, the mesh filter must be designed at a suitable cutoff frequency to allow the passage of the frequency variation. The VCO will monitor the frequency variation of the modulated carrier and the error signal give access to the modulating signal.

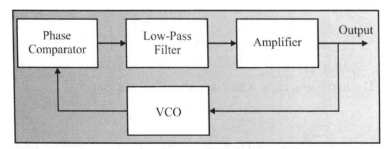

Figure 11.32 Block diagram of the PLL in the FM demodulation. In this case, the output of PLL is in the amplifier.

As mentioned, the output frequency of the VCO signal is set to be equal to the center frequency of the FM carrier. In this case, the signals are in synchronism and the phase error between the signals is null or constant.

The message signal transmitted by the FM wave is then obtained when, in varying the frequency of the FM carrier, the new comparison with the VCO signal, which is tuned to the frequency of the FM wave, presents a new lag signal which, being the input of the mesh filter, results in a voltage representing the message signal.

In this case, each frequency of the FM carrier is related to an average voltage obtained at the output of the filter. If the frequency of the FM wave increases beyond the center frequency, the average voltage at the output of the filter also increases compared to that obtained when the signals are synchronized. On the other hand, when the frequency of the FM carrier decreases, the output voltage of the mesh filter also decreases. This is the characteristic process of FM modulation. In this way, the message signal, transmitted by the frequency variation of the FM carrier, is then obtained at the output of the low pass filter (Mayaram, 2008).

11.6.2 Frequency Discriminator

The frequency discriminator is a device that has the purpose of converting frequencies into amplitude variations so that when applying an FM modulated signal to its input, its output is a linear voltage proportional to the input frequency.

To demodulate the signal, a frequency discriminator uses an amplitude limiter, a bandpass filter, a discriminator, and an envelope detector, as illustrated in Figure 11.33.

The amplitude limiter receives the FM modulated signal and eliminates signal amplitude variations. The bandpass filter, which represents a tuned circuit, has in its frequency response an approximately linear region in which the discriminator operates. The frequency of the tuned filter should not match the frequency of the FM signal. It must be centered on the frequency spectrum and the frequency of the FM signal must be located in the approximately

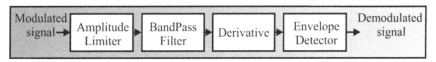

Figure 11.33 Block diagram of a frequency discriminator.

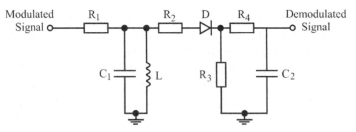

Figure 11.34 Electrical circuit of a frequency discriminator.

linear regions of the spectrum, that is, above or below the resonant frequency. The envelope detector receives an AM signal from the discriminator and can thus detect it.

A feature of the discriminator is that the FM signal must have a small frequency shift so that the demodulation process best fits the linear region of the tuned filter's frequency spectrum. Figure 11.34 illustrates the electrical diagram of a practical FM demodulator with the frequency discriminator (Gomes, 1985).

Appendix A: Fourier Theory

This chapter presents historical aspects of the development of the Fourier theory, along with details of his life and work. It also provides the mathematical basis for the reader to understand the Fourier theory. There is an introduction to signal analysis. The main concepts associated with the Fourier series and Fourier transform are introduced. The theory and the properties of the Fourier transform are presented, along with the main properties of signal and system analysis.

A.1 Introduction

This chapter provides the necessary mathematical basis, and some historical context, for the reader to understand the Fourier Theory (Alencar and Rocha Jr., 2020) (Baskakov, 1986). The main concepts associated with the Fourier series and Fourier transform are introduced. The theory and the properties of the Fourier transform are presented, along with the main properties of signal and system analysis (Papoulis, 1983).

A.2 The Concept of Integration

The Fourier theory establishes fundamental conditions for the representation of an arbitrary function in a finite interval as a sum of sines and cosines, in a finite interval. As a matter of fact, this is just a simplified version of the general Fourier representation of signals in which a periodic signal $f(t)$ can be represented by a complete set of orthogonal functions (Fourier, 1888).

The periodic signal $f(t)$ need to satisfy the Dirichlet conditions, that is, $f(t)$ is a bounded function which has at most a finite number of local maxima and minima and a finite number of discontinuity points, in any period of time (Wylie, 1966).

The representation of signals by orthogonal functions usually presents an error, which decreases as the number of component terms in the corresponding series is increased. This error appears as the Gibbs phenomenon, an oscillation that occurs at the transition points (Schwartz and Shaw, 1975).

The familiar Riemann integral applies to continuous functions that do not present too many points of discontinuity. Therefore, it is not adequate to use the Riemann integral for a generic measurable function, for example. In fact, a function f can be discontinuous everywhere, or, maybe, there is no sense in saying that a function is continuous if it is defined for an abstract set (Kolmogorov and Formin, 1970).

Bernhard Riemann (1826–1866) developed the definite integral that bears his name to establish with rigor the exhaustion method, created by Eudoxus of Cnido (390–338 BC), and later improved by Archimedes (287– 212 BC). Isaac Newton (1642–1727) created the Method of Fluxions, to explain the derivative, and Gottfried Wilhelm Leibniz (1646–1716) gave Calculus its present form.

The first rigorous treatment of the integral was given by Augustin Louis Cauchy (1789–1857), but Jean-Baptiste Joseph Fourier (1768–1830) noticed that the definition in terms of the derivative was too restrictive.

In 1907, Fourier published an article on the diffusion of heat, in which he discussed the representation of sets of functions by trigonometric series, and managed to prove that the solution of the equation

$$f(x) = a_0 + \sum_{n=1}^{\infty} a_n \cos nx + b_n \sin nx, \tag{A.1}$$

for an infinite number of unknown coefficients, could be expressed in terms of integrals. He also proved that the solution exists even for discontinuous functions, whose antiderivatives cannot be found.

Riemann submitted his *Habilitationsschrift* (probationary essay) for admission to the faculty in 1853, entitled "On the Representation of a Function by Means of a Trigonometrical Series", in which he extended the Cauchy integral to a larger class of functions, and established the necessary and sufficient conditions to represent a function by a Fourier series (Phillips, 1984).

A.3 Basic Fourier Analysis

The expansion of a periodic signal $f(t)$ as a sum of mutually orthogonal functions requires a review of the concepts of periodicity and orthogonality. A given function $f(t)$ is periodic, of period T, as illustrated in Figure A.1, if

A.3 Basic Fourier Analysis 221

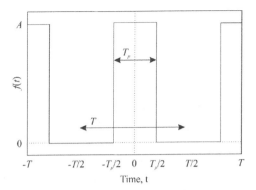

Figure A.1 Example of a periodic signal.

and only if, T is the smallest positive number for which $f(t + T) = f(t)$. In other words, $f(t)$ is periodic if its domain contains $t + T$ whenever it contains t, and $f(t + T) = f(t)$.

It follows from the definition of a periodic function that if T represents the period of $f(t)$, then $f(t) = f(t + nT)$, for $n = 1, 2, \ldots$, that is, $f(t)$ repeats its values when integer multiples of T (Wozencraft and Jacobs, 1965) are added to its argument.

If $f(t)$ and $g(t)$ are two periodic functions with the same period T, then their sum $f(t) + g(t)$ will also be a periodic function with period T. This result can be proven if one makes $h(t) = f(t) + g(t)$, and notices that $h(t + T) = f(t + T) + g(t + T) = f(t) + g(t) = h(t)$.

Orthogonality provides the tool to introduce the concept of a basis, that is, a minimum set of functions that can be used to generate other functions. However, orthogonality by itself does not guarantee that a complete vector space is generated.

Two real functions $f(t)$ and $g(t)$, defined in the interval $-T/2 \leq t \leq T/2$, are orthogonal if their inner product is null, that is, if

$$< f(t), g(t) > = \int_{-T/2}^{T/2} f(t)g(t)dt = 0. \quad (A.2)$$

A.3.1 The Trigonometric Fourier Series

The trigonometric Fourier series representation of a signal $f(t)$ can be written as

$$f(t) = a_0 + \sum_{n=1}^{\infty} [a_n \cos(n\omega_0 t) + b_n \sin(n\omega_0 t)], \quad (A.3)$$

in which the term a_0 represents the average value of the function $f(t)$, ω_0 is the angular frequency, a_n and b_n are the Fourier series coefficients, in which n is a positive integer. The equality sign holds in (A.3) for all values of t only when $f(t)$ is periodic.

Although it has been initially proposed by Fourier to solve heat diffusion problems, the Fourier series representation is useful for any type of signal, as long as that signal representation is required only in the $[0, T]$ interval. Outside that interval, the Fourier series representation is periodic, even if the signal $f(t)$ is not periodic (Knopp, 1990).

The trigonometric functions because satisfy the following equations, called orthogonality relations, for integer values of n and m:

$$\int_0^T \cos(n\omega_0 t)\sin(m\omega_0 t)dt = 0, \quad \text{for all integers } n, m, \tag{A.4}$$

$$\int_0^T \cos(n\omega_0 t)\cos(m\omega_0 t)dt = \begin{cases} 0 & \text{if } n \neq m \\ \frac{T}{2} & \text{if } n = m \end{cases} \tag{A.5}$$

$$\int_0^T \sin(n\omega_0 t)\sin(m\omega_0 t)dt = \begin{cases} 0 & \text{if } n \neq m \\ \frac{T}{2} & \text{if } n = m \end{cases} \tag{A.6}$$

in which $\omega_0 = 2\pi/T$.

Explicit expressions for the coefficients a_n and b_n of the Fourier trigonometric series can be computed, integrating both sides in expression (A.3) in the interval $[0, T]$, as shown in the following (Oberhettinger, 1990)

$$\int_0^T f(t)dt = \int_0^T a_o dt + \sum_{n=1}^{\infty} \int_0^T a_n \cos(n\omega_0 t)dt$$

$$+ \sum_{n=1}^{\infty} \int_0^T b_n \sin(n\omega_0 t)dt$$

and since

$$\int_0^T a_n \cos(n\omega_0 t)dt = \int_0^T b_n \sin(n\omega_0 t)dt = 0,$$

it follows that

$$a_o = \frac{1}{T}\int_0^T f(t)dt. \tag{A.7}$$

A.3 Basic Fourier Analysis

Multiplication of both sides in expression (A.3) by $\cos(m\omega_0 t)$, and integrating in the interval $[0, T]$, leads to

$$\int_0^T f(t) \cos(m\omega_0 t) dt = \int_0^T a_o \cos(m\omega_0 t) dt$$

$$+ \sum_{n=1}^{\infty} \int_0^T a_n \cos(n\omega_0 t) \cos(m\omega_0 t) dt \quad (A.8)$$

$$+ \sum_{n=1}^{\infty} \int_0^T b_n \cos(m\omega_0 t) \sin(n\omega_0 t) dt,$$

which, after simplification, gives

$$a_n = \frac{2}{T} \int_0^T f(t) \cos(n\omega_0 t) dt, \quad (A.9)$$

for $n = 1, 2, 3, \ldots$.

The coefficient b_n is found multiplying both sides in expression (A.3) by $\sin(n\omega_0 t)$ and integrating in the interval $[0, T]$, that is,

$$b_n = \frac{2}{T} \int_0^T f(t) \sin(n\omega_0 t) dt, \quad (A.10)$$

for $n = 1, 2, 3, \ldots$.

It is possible to simplify the computation of coefficients of a trigonometric Fourier series, using properties of even and odd functions.

If $f(t)$ is an even function, then $b_n = 0$, and

$$a_n = \frac{2}{T} \int_0^T f(t) \cos(n\omega_0 t) dt, \quad (A.11)$$

for $n = 1, 2, 3, \ldots$.

If $f(t)$ is an odd function, then $a_n = 0$ and

$$b_n = \frac{2}{T} \int_0^T f(t) \sin(n\omega_0 t) dt, \quad (A.12)$$

for $n = 1, 2, 3, \ldots$.

Example: Determine the coefficients of the trigonometric Fourier series for the signal $f(t) = [u(\cos(2\pi f t))]$, which has period $T = 1/f$, in which $u(t)$ denotes the Heaviside unit step function (Alencar, Rocha Jr., 2005).

224 Appendix A: Fourier Theory

Solution: Because of the signal symmetry, with respect to the ordinate axis, it follows that $f(t) = f(-t)$. Therefore, $b_n = 0$, and all that is left to compute is a_o, and a_n for $n = 1, 2, \ldots$.

The expression to calculate the average value a_o is

$$a_o = \frac{1}{T} \int_{-\frac{T}{2}}^{\frac{T}{2}} f(t) dt = \frac{1}{T} \int_{-\tau}^{\tau} dt = \frac{2\tau}{T}.$$

In the previous equation, the maximum value of τ is $T/2$. The coefficients a_n for $n = 1, 2, \ldots$ are computed as

$$a_n = \frac{2}{T} \int_0^T f(t) \cos(n\omega_0 t) dt = \frac{2}{T} \int_{-\tau}^{\tau} \cos(n\omega_0 t) dt,$$

$$a_n = \frac{4}{T} \int_0^\tau \cos(n\omega_0 t) dt = \frac{4}{T n \omega_0} \sin(n\omega_0 t) \Big|_0^\tau = \left(\frac{4\tau}{T}\right) \frac{\sin(n\omega_0 \tau)}{n\omega_0 \tau}.$$

Therefore, the signal $f(t)$ is represented by the following trigonometric Fourier series

$$f(t) = \frac{2\tau}{T} + \left(\frac{4\tau}{T}\right) \sum_{n=1}^{\infty} \frac{\sin(n\omega_0 \tau)}{n\omega_0 \tau} \cos(n\omega_0 t).$$

A.3.2 The Compact Fourier Series

The compact Fourier series is another manner to represent the Fourier series, as follows

$$f(t) = C_0 + \sum_{n=1}^{\infty} C_n \cos(n\omega_0 t + \theta_n). \tag{A.13}$$

Expanding the expression $C_n \cos(n\omega_0 t + \theta)$ as $C_n \cos(n\omega_0 t) + \cos\theta_n - C_n \sin(n\omega_0 t) \sin\theta_n$ and comparing this result with (A.3), it follows that $a_o = C_0$, $a_n = C_n \cos\theta_n$, and $b_n = -C_n \sin\theta_n$. It is now possible to compute C_n as a function of a_n and b_n. For that purpose, it is sufficient to square a_n and b_n and add the result, that is,

$$a_n^2 + b_n^2 = C_n^2 \cos^2\theta_n + C_n^2 \sin^2\theta_n = C_n^2. \tag{A.14}$$

From Equation 14, the modulus of C_n can be written as

$$C_n = \sqrt{a_n^2 + b_n^2}. \tag{A.15}$$

In order to determine θ_n, it suffices to divide b_n by a_n, that is,

$$\frac{b_n}{a_n} = -\frac{\sin\theta_n}{\cos\theta_n} = -\tan\theta_n, \tag{A.16}$$

which, when solved for θ_n, gives

$$\theta_n = -\arctan\left(\frac{b_n}{a_n}\right). \tag{A.17}$$

A.3.3 The Exponential Fourier Series

Since the set of exponential functions $e^{jn\omega_0 t}$, $n = 0, \pm 1, \pm 2, \ldots$, is a complete set of complex orthogonal functions in an interval of magnitude T, in which $T = 2\pi/\omega_0$, then it is possible to represent a function $f(t)$ by a linear combination of exponential functions in an interval T.

$$f(t) = \sum_{-\infty}^{\infty} F_n e^{jn\omega_0 t} \tag{A.18}$$

in which

$$F_n = \frac{1}{T} \int_{-\frac{T}{2}}^{\frac{T}{2}} f(t) e^{-jn\omega_0 t} dt. \tag{A.19}$$

Formula (A.18) represents the exponential Fourier series expansion of $f(t)$ and Formula 19 permits to compute the associated series coefficients. The exponential Fourier series is also known as the complex Fourier series. It can be shown that Equation 18 is just another way of expressing the Fourier series as given in (A.3). Using Eulert's identity, replacing $\cos(n\omega_0 t) + j\sin(n\omega_0 t)$ for $e^{n\omega_0 t}$ in (A.18), it follows that

$$f(t) = F_o + \sum_{n=-\infty}^{-1} F_n[\cos(n\omega_0 t) + j\sin(n\omega_0 t)]$$

$$+ \sum_{n=1}^{\infty} F_n[\cos(n\omega_0 t) + j\sin(n\omega_0 t)],$$

or

$$f(t) = F_o + \sum_{n=1}^{\infty} F_n[\cos(n\omega_0 t) + j\sin(n\omega_0 t)]$$

$$+ F_{-n}[\cos(n\omega_0 t) - j\sin(n\omega_0 t)].$$

Appendix A: Fourier Theory

After grouping the coefficients of the sine and cosine terms, one obtains

$$f(t) = F_o + \sum_{n=1}^{\infty}(F_n + F_{-n})\cos(n\omega_0 t) + j(F_n - F_{-n})\sin(n\omega_0 t). \quad (A.20)$$

Comparing this expression with Expression 3 it follows that

$$a_o = F_0, \quad a_n = (F_n - F_{-n}) \quad \text{and} \quad b_n = j(F_n - F_{-n}), \quad (A.21)$$

and that

$$F_o = a_o, \quad (A.22)$$

$$F_n = \frac{a_n - jb_n}{2}, \quad (A.23)$$

and

$$F_{-n} = \frac{a_n + jb_n}{2}. \quad (A.24)$$

If the function $f(t)$ is even, that is, if $b_n = 0$, then

$$a_o = F_0, \quad F_n = \frac{a_n}{2}, \quad \text{and} \quad F_{-n} = \frac{a_n}{2}. \quad (A.25)$$

Example: Compute the exponential Fourier series for the train of impulses given by,

$$\delta_T(t) = \delta[\sin(2\pi f t)], \quad f = 1/T.$$

Solution: The complex coefficients are given by

$$F_n = \frac{1}{T}\int_{\frac{-T}{2}}^{\frac{T}{2}} \delta_T(t) e^{-jn\omega_0 t} dt = \frac{1}{T}, \quad (A.26)$$

using the property of impulse filtering,

$$\int_{-\infty}^{\infty} \delta(t - t_o) f(t) dt = f(t_o). \quad (A.27)$$

On the other hand, $f(t)$ can be written as

$$f(t) = \frac{1}{T}\sum_{n=-\infty}^{\infty} e^{-jn\omega_0 t}. \quad (A.28)$$

A.4 Fourier Transform   227

The impulse train, as well as, the impulse function itself, are idealizations, as most functions are, of real signals. It is obtained by passing a digital signal through a differentiator circuit and then passing the resulting waveform through a half-wave rectifier.

According to its properties, the Fourier series expansion of a periodic signal is equivalent to its decomposition in frequency components. In general, a periodic function with period T has frequency components $0, \pm \omega_0, \pm 2\omega_0, \pm 3\omega_0, \ldots, \pm n\omega_0$, in which $\omega_0 = 2\pi/T$ is the fundamental frequency and the multiples of ω_0 are called harmonics. In this case, the spectrum exists only for discrete values of ω and that the spectral components are spaced by at least ω_0.

A.4 Fourier Transform

An arbitrary function can be represented in terms of an exponential, or trigonometric, Fourier series in a finite interval. If such a function is periodic, this representation can be extended for the entire interval $(-\infty, \infty)$.

However, it is interesting to observe the spectral behavior of a function in general, periodic or not, in the entire interval $(-\infty, \infty)$. To do that, it is necessary to truncate the function $f(t)$ in the interval $[-T/2, T/2]$, to obtain $f_T(t)$. It is possible then to represent this function as a sum of exponentials in the entire interval $(-\infty, \infty)$ if T goes to infinity, as follows.

$$lim_{T \to \infty} f_T(t) = f(t).$$

The truncated signal $f_T(t)$ can be represented by an exponential Fourier series as

$$f_T(t) = \sum_{n=-\infty}^{\infty} F_n e^{jn\omega_0 t}, \qquad (A.29)$$

in which $\omega_0 = 2\pi/T$ and

$$F_n = \frac{1}{T} \int_{-\frac{T}{2}}^{\frac{T}{2}} f_T(t) e^{-jn\omega_0 t} dt. \qquad (A.30)$$

The complex coefficients F_n represent the spectral amplitude associated with each component of frequency $n\omega_0$. As the interval T increases, the amplitudes diminish but the spectrum shape is not altered. The increase in T forces ω_0 to diminish and the spectrum to become denser. In the limit, as $T \to \infty$, ω_0 becomes infinitesimally small, being represented by $d\omega$. There

Appendix A: Fourier Theory

are now infinitely many components and the spectrum is no longer a discrete one, becoming a continuous spectrum in the limit.

For convenience, write $TF_n = F(\omega)$, that is, the product TF_n becomes a function of the variable ω, since $n\omega_0 \to \omega$. Replacing $\frac{F(\omega)}{T}$ for F_n in 29, one obtains

$$f_T(t) = \frac{1}{T} \sum_{n=-\infty}^{\infty} F(\omega)e^{j\omega t}. \tag{A.31}$$

Replacing $\omega_0/2\pi$ for $1/T$,

$$f_T(t) = \frac{1}{2\pi} \sum_{n=-\infty}^{\infty} F(\omega)e^{j\omega t}\omega_0. \tag{A.32}$$

In the limit, as T approaches infinity, one has

$$f(t) = \frac{1}{2\pi} \int_{-\infty}^{\infty} F(\omega)e^{j\omega t}d\omega \tag{A.33}$$

which is known as the inverse Fourier transform.

Similarly, from (A.30), as T approaches infinity, one obtains

$$F(\omega) = \int_{-\infty}^{\infty} f(t)e^{-j\omega t}dt \tag{A.34}$$

which is known as the direct Fourier transform, sometimes denoted in the literature as $F(\omega) = \mathcal{F}[f(t)]$. A Fourier transform pair is frequently denoted as $f(t) \leftrightarrow F(\omega)$. Some important Fourier transforms are presented in the following sections (Haykin, 1988).

A.4.1 Bilateral Exponential Signal

The bilateral exponential signal, also called the Laplace function, $f(t) = e^{-a|t|}$, frequently appears in probability stochastic processes analyses. Its Fourier transform can be computed directly from Formula 34,

$$F(\omega) = \int_{-\infty}^{\infty} e^{-a|t|}e^{-j\omega t}dt$$

$$= \int_{-\infty}^{0} e^{at}e^{-j\omega t}dt + \int_{0}^{\infty} e^{-at}e^{-j\omega t}dt \tag{A.35}$$

$$= \frac{1}{a-j\omega} + \frac{1}{a+j\omega},$$
$$F(\omega) = \frac{2a}{a^2+\omega^2}. \quad (A.36)$$

A.4.2 Transform of the Gate Function

A gate function, $u(t) - u(t-T)$, is useful to establish the domain of a signal. The Fourier transform of this function can be calculated as follows,

$$\begin{aligned} F(\omega) &= \int_{-\frac{T}{2}}^{\frac{T}{2}} e^{-j\omega t} dt \\ &= \frac{1}{j\omega}(e^{j\omega \frac{T}{2}} - e^{-j\omega \frac{T}{2}}) \quad (A.37) \\ &= \frac{1}{j\omega} 2j \sin(\omega T/2), \end{aligned}$$

which can be rearranged as

$$F(\omega) = T\left(\frac{\sin(\omega T/2)}{\omega T/2}\right),$$

and finally

$$F(\omega) = T\text{Sa}\left(\frac{\omega T}{2}\right), \quad (A.38)$$

in which $Sa(x) = \frac{\sin x}{x}$ is the sampling function. This function converges to one, as x goes to zero. The sampling function, illustrated in Figure A.2, is very useful in communication theory.

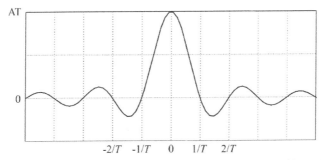

Figure A.2 Graphic of the Fourier transform of the gate function $u(t) - u(t-T)$.

A.4.3 Fourier Transform of the Impulse Function

Making the substitution, $f(t) = \delta(t)$ in Formula 34, using the impulse filtering property, one obtains

$$F(\omega) = \int_{-\infty}^{\infty} \delta(t) e^{-j\omega t} dt = 1, \quad (A.39)$$

Because the Fourier transform is constant, over the entire spectrum, one concludes that the impulse function contains a continuum of equal amplitude spectral components.

An impulse can model a sudden electrostatic discharge, such as the lightning that occurs during a thunderstorm. The effect of the lightning can be felt, for example, in several radiofrequency ranges, including the AM and FM radio bands, and the TV band.

Alternatively, by making $F(\omega) = 1$ in (A.33) and simplifying, the impulse function can be written as

$$\delta(t) = \frac{1}{\pi} \int_0^{\infty} \cos \omega t \, d\omega.$$

A.4.4 Transform of the Constant Function

If $f(t)$ is a constant function, then its Fourier transform in principle would not exist, since this function does not satisfy the absolute integrability criterion. In general, $F(\omega)$, the Fourier transform of $f(t)$, is expected to be finite, that is,

$$|F(\omega)| \leq \int_{-\infty}^{\infty} |f(t)||e^{-j\omega t}| dt < \infty, \quad (A.40)$$

since $|e^{-j\omega t}| = 1$, then

$$\int_{-\infty}^{\infty} |f(t)| dt < \infty. \quad (A.41)$$

However, that is just a sufficiency condition and not a necessary condition for the existence of the Fourier transform, since there exist functions that although do not satisfy the condition of absolute integrability, in the limit do have a Fourier transform (Carlson, 1975).

This approach is often used to compute Fourier transforms of several functions. For instance, the constant function can be approximated by a gate

A.4 Fourier Transform

function with amplitude A and width τ, if τ approaches the infinity,

$$\mathcal{F}[A] = \lim_{\tau \to \infty} A\tau \text{Sa}\left(\frac{\omega\tau}{2}\right)$$
$$= 2\pi A \lim_{\tau \to \infty} \frac{\tau}{2\pi} \text{Sa}\left(\frac{\omega\tau}{2}\right) \tag{A.42}$$
$$\mathcal{F}[A] = 2\pi A \delta(\omega). \tag{A.43}$$

This result is intuitive because a constant function in time represents a DC level and contains no spectral component except for the one at $\omega = 0$.

A.4.5 Fourier Transform of the Sine and Cosine Function

The sine and the cosine functions are periodic functions, therefore, they do not satisfy the condition of absolute integrability. However, their respective Fourier transforms exist in the limit when τ goes to infinity. Assuming the function to exist only in the interval $\left(\frac{-\tau}{2}, \frac{\tau}{2}\right)$ and to be zero outside this interval, and considering the limits of the expression when τ goes to infinity,

$$\mathcal{F}(\sin \omega_0 t) = \lim_{\tau \to \infty} \int_{\frac{-\tau}{2}}^{\frac{\tau}{2}} \sin \omega_0 t e^{-j\omega t} dt$$

$$= \lim_{\tau \to \infty} \int_{\frac{-\tau}{2}}^{\frac{\tau}{2}} \frac{e^{-j(\omega - \omega_0)t}}{2j} - \frac{e^{-j(\omega + \omega_0)t}}{2j} dt \tag{A.44}$$

$$= \lim_{\tau \to \infty} \left[\frac{j\tau \sin(\omega + \omega_0)\frac{\tau}{2}}{2(\omega + \omega_0)\frac{\tau}{2}} - \frac{j\tau \sin(\omega - \omega_0)\frac{\tau}{2}}{2(\omega - \omega_0)\frac{\tau}{2}} \right]$$

$$= \lim_{\tau \to \infty} \left\{ j\frac{\tau}{2} \text{Sa}\left[\frac{(\omega + \omega_0)}{2}\right] - j\frac{\tau}{2} \text{Sa}\left[\frac{\tau(\omega + \omega_0)}{2}\right] \right\}.$$

Therefore, $\mathcal{F}(\sin \omega_0 t) = j\pi[\delta(\omega + \omega_0) - \delta(\omega - \omega_0)]$.
Applying similar reasoning, it follows that

$$\mathcal{F}(\cos \omega_0 t) = \pi[\delta(\omega - \omega_0) + \delta(\omega + \omega_0)], \tag{A.45}$$

which is shown in Figure A.3.

A.4.6 Fourier Transform of the Complex Exponential

The Fourier transform can be obtained using Euler's identity, $e^{j\omega_0 t} = \cos \omega_0 t + j \sin \omega_0 t$, and a property, as follows

$$\mathcal{F}[e^{j\omega_0 t}] = \mathcal{F}[\cos \omega_0 t + j \sin \omega_0 t]. \tag{A.46}$$

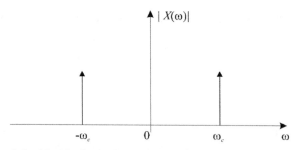

Figure A.3 Magnitude plot for the Fourier transform of the sine function.

Substituting in (A.46) the Fourier transforms of the sine and of the cosine functions, respectively, it follows that

$$\mathcal{F}[e^{j\omega_0 t}] = 2\pi\delta(\omega - \omega_0). \tag{A.47}$$

A.4.7 Fourier Transform of a General Periodic Function

Consider the exponential Fourier series representation of a general periodic function $f_T(t)$ of period T,

$$f_T(t) = \sum_{n=-\infty}^{\infty} F_n e^{jn\omega_0 t}. \tag{A.48}$$

Applying the Fourier transform Fourier to both sides in Equation 50, it follows that

$$\mathcal{F}[f_T(t)] = \mathcal{F}\left[\sum_{n=-\infty}^{\infty} F_n e^{jn\omega_0 t}\right] \tag{A.49}$$

$$= \sum_{n=-\infty}^{\infty} F_n \mathcal{F}[e^{jn\omega_0 t}]. \tag{A.50}$$

Applying the result from (A.47) in (A.50), it follows that

$$F(\omega) = 2\pi \sum_{n=-\infty}^{\infty} F_n \delta(\omega - n\omega_0). \tag{A.51}$$

A.5 Properties of the Fourier Transform

As the Fourier integral can be sometimes difficult to solve for certain functions, it is important to obtain a set of properties to help solve more complex problems in signal analysis.

A.5.1 Linearity of the Fourier Transform

The Fourier transform is a linear operator, that is, if a function can be written as a linear combination of other functions, the corresponding Fourier transform will be given by a linear combination of the corresponding Fourier transforms of each one of the functions involved in the linear combination (Gagliardi, 1988).

If $f(t) \leftrightarrow F(\omega)$ and $g(t) \leftrightarrow G(\omega)$, it then follows that

$$\alpha f(t) + \beta g(t) \longleftrightarrow \alpha F(\omega) + \beta G(\omega). \tag{A.52}$$

Proof: Let $h(t) = \alpha f(t) + \beta g(t) \rightarrow$, then it follows that

$$H(\omega) = \int_{-\infty}^{\infty} h(t) e^{-j\omega t} dt$$

$$= \alpha \int_{-\infty}^{\infty} f(t) e^{-j\omega t} dt + \beta \int_{-\infty}^{\infty} g(t) e^{-j\omega t} dt,$$

and finally

$$H(\omega) = \alpha F(\omega) + \beta G(\omega). \tag{A.53}$$

A.5.2 Scaling Property

In case a function $f(t)$ is scaled by a parameter, to obtain $f(\alpha t)$, its Fourier transform is also scaled by the same value, as shown in the following.

$$\mathcal{F}[f(\alpha t)] = \int_{-\infty}^{\infty} f(\alpha t) e^{-j\omega t} dt. \tag{A.54}$$

Initially, consider $\alpha > 0$ in (A.54). By letting $u = \alpha t$, it follows that $dt = (1/\alpha) du$. Replacing u for αt in (A.54), one obtains

$$\mathcal{F}[f(\alpha t)] = \int_{-\infty}^{\infty} \frac{f(u)}{\alpha} e^{-j\frac{\omega}{\alpha} u} du$$

which simplifies to
$$\mathcal{F}[f(\alpha t)] = \frac{1}{\alpha} F\left(\frac{\omega}{\alpha}\right).$$

Consider the case in which $\alpha < 0$. By a similar procedure, it follows that
$$\mathcal{F}[f(\alpha t)] = -\frac{1}{\alpha} F\left(\frac{\omega}{\alpha}\right).$$

Therefore, finally
$$\mathcal{F}[f(\alpha t)] = \frac{1}{|\alpha|} F\left(\frac{\omega}{\alpha}\right). \tag{A.55}$$

This result indicates that if a signal is compressed in the time domain by a factor α, then its Fourier transform will expand in the frequency domain by the same factor.

A.5.3 Symmetry Property

This property is important because it can be used to compute half of all the other Fourier transform properties. The symmetry property states that if
$$f(t) \longleftrightarrow F(\omega), \tag{A.56}$$

then it follows that
$$F(t) \longleftrightarrow 2\pi f(-\omega). \tag{A.57}$$

Proof: By definition,
$$f(t) = \frac{1}{2\pi} \int_{-\infty}^{+\infty} F(\omega) e^{j\omega t} d\omega,$$

which after multiplication of both sides by 2π becomes
$$2\pi f(t) = \int_{-\infty}^{+\infty} F(\omega) e^{j\omega t} d\omega.$$

By letting $u = -t$, it follows that
$$2\pi f(-u) = \int_{-\infty}^{+\infty} F(\omega) e^{-j\omega u} d\omega,$$

and now by making $t = \omega$, one obtains
$$2\pi f(-u) = \int_{-\infty}^{+\infty} F(t) e^{-jtu} dt.$$

Finally, by letting $u = \omega$, it follows that

$$2\pi f(-\omega) = \int_{-\infty}^{+\infty} F(t)e^{-j\omega t}dt. \tag{A.58}$$

A.5.4 Time Domain Shift

Given a function and its transform, $f(t) \leftrightarrow F(\omega)$, it then follows that $f(t - \sigma) \leftrightarrow F(\omega)e^{-j\omega\sigma}$. Let $g(t) = f(t - \sigma)$. In this case, it follows that

$$G(\omega) = \mathcal{F}[g(t)] = \int_{-\infty}^{\infty} f(t - \sigma)e^{-j\omega t}dt. \tag{A.59}$$

By making $\tau = t - \sigma$ it follows that

$$G(\omega) = \int_{-\infty}^{\infty} f(\tau)e^{-j\omega(\tau+\sigma)}d\tau \tag{A.60}$$

$$= \int_{-\infty}^{\infty} f(\tau)e^{-j\omega\tau}e^{-j\omega\sigma}d\tau, \tag{A.61}$$

and finally

$$G(\omega) = e^{-j\omega\sigma}F(\omega). \tag{A.62}$$

This result shows that if a function is shifted in time, its frequency domain amplitude spectrum remains the same. However, the corresponding phase spectrum suffers a rotation proportional to $\omega\sigma$.

A.5.5 Frequency Domain Shift

Given that $f(t) \leftrightarrow F(\omega)$, it then follows that $f(t)e^{j\omega_0 t} \leftrightarrow F(\omega - \omega_0)$.

$$\mathcal{F}[f(t)e^{j\omega_0 t}] = \int_{-\infty}^{\infty} f(t)e^{j\omega_0 t}e^{-j\omega t}dt$$

$$= \int_{-\infty}^{\infty} f(t)e^{-j(\omega-\omega_0)t}dt, \tag{A.63}$$

$$\mathcal{F}[f(t)e^{j\omega_0 t}] = F(\omega - \omega_0). \tag{A.64}$$

A.5.6 Differentiation in the Time Domain

The differentiation property is useful to obtain the Fourier transform of many generalized functions. Given that

$$f(t) \longleftrightarrow F(\omega), \tag{A.65}$$

it follows that
$$\frac{df(t)}{dt} \longleftrightarrow j\omega F(\omega). \tag{A.66}$$

Proof: Consider the expression for the inverse Fourier transform
$$f(t) = \frac{1}{2\pi} \int_{-\infty}^{\infty} F(\omega) e^{j\omega t} d\omega. \tag{A.67}$$

Differentiating in time, it follows that
$$\begin{aligned}\frac{df(t)}{dt} &= \frac{1}{2\pi} \frac{\partial}{\partial t} \int_{-\infty}^{\infty} F(\omega) e^{j\omega t} d\omega \\ &= \frac{1}{2\pi} \int_{-\infty}^{\infty} F(\omega) \frac{\partial}{\partial t} e^{j\omega t} d\omega \\ &= \frac{1}{2\pi} \int_{-\infty}^{\infty} j\omega F(\omega) e^{j\omega t} d\omega,\end{aligned}$$

and then
$$\frac{df(t)}{dt} \longleftrightarrow j\omega F(\omega). \tag{A.68}$$

In the general case, after differentiating n times, it follows that
$$\frac{d^n f(t)}{dt} \longleftrightarrow (j\omega)^n f(\omega). \tag{A.69}$$

Using properties of the Fourier transform it can be shown that the derivative of the signal $h(t) = f(t) * g(t)$ can be written as
$$h'(t) = f'(t) * g(t), \quad \text{or} \quad h'(t) = f(t) * g'(t).$$

A.5.7 Integration in the Time Domain

Suppose $f(t)$ is a signal with zero mean value, that is, let $\int_{-\infty}^{\infty} f(t)dt = 0$. By defining
$$g(t) = \int_{-\infty}^{t} f(\tau)d\tau, \tag{A.70}$$

it follows that
$$\frac{dg(t)}{dt} = f(t),$$

and since
$$g(t) \longleftrightarrow G(\omega), \tag{A.71}$$

A.5 Properties of the Fourier Transform

then
$$f(t) \longleftrightarrow j\omega G(\omega),$$

and
$$G(\omega) = \frac{F(\omega)}{j\omega}. \tag{A.72}$$

In this manner, it follows that for a signal with zero average value

$$f(t) \longleftrightarrow F(\omega)$$

$$\int_{-\infty}^{t} f(\tau) d\tau \longleftrightarrow \frac{F(\omega)}{j\omega}. \tag{A.73}$$

Generalizing, for the case in which $f(t)$ has a non-zero average value, it follows that

$$\int_{-\infty}^{t} f(\tau) d\tau \longleftrightarrow \frac{F(\omega)}{j\omega} + \pi\delta(\omega)F(0). \tag{A.74}$$

Using the property, and recalling that $\delta(t) \leftrightarrow 1$, it can be shown that the Fourier transform of the unit step function is given by

$$u(t) \longleftrightarrow \frac{1}{j\omega} + \pi\delta(\omega).$$

A.5.8 Convolution Theorem in the Time Domain

The convolution theorem can be used to obtain the response of a linear system to an input signal. It can be also used to obtain the sampling theorem, which is discussed in the next section.

Let $h(t) = f(t) * g(t)$ and let $h(t) \leftrightarrow H(\omega)$. It follows that

$$H(\omega) = \int_{-\infty}^{\infty} h(t) e^{-j\omega t} dt = \int_{-\infty}^{\infty} \int_{-\infty}^{\infty} f(\tau) g(t-\tau) e^{-j\omega t} dt d\tau. \tag{A.75}$$

$$H(\omega) = \int_{-\infty}^{\infty} f(\tau) \int_{-\infty}^{\infty} g(t-\tau) e^{-j\omega t} dt d\tau, \tag{A.76}$$

$$H(\omega) = \int_{-\infty}^{\infty} f(\tau) G(\omega) e^{-j\omega \tau} d\tau \tag{A.77}$$

and finally,
$$H(\omega) = F(\omega) G(\omega). \tag{A.78}$$

The convolution of two-time functions is equivalent in the frequency domain to the product of their respective Fourier transforms.

A.5.9 Convolution Theorem in the Frequency Domain

If a function in time is the product of two other functions, $h(t) = f(t) \cdot g(t)$, it is possible to obtain the Fourier transform proceeding in a way similar to the previous derivation.

$$H(\omega) = \frac{1}{2\pi}[F(\omega) * G(\omega)]. \tag{A.79}$$

The product of two-time functions has a Fourier transform given by the convolution of their respective Fourier transforms. The convolution operation is often used when computing the response of a linear circuit, given its impulse response and an input signal.

Example: Using the frequency convolution theorem, it can be shown that if the cosine signal is switched on at zero time, the resulting Fourier transform is given by

$$\cos(\omega_c t)u(t) \longleftrightarrow \frac{\pi}{2}[\delta(\omega + \omega_c) + \delta(\omega - \omega_c)] + j\frac{\omega}{\omega_c^2 - \omega^2}.$$

A.6 Sampling Theorem

A band-limited signal $f(t)$, that has no frequency components above $\omega_M = 2\pi f_M$, can be reconstructed from its samples, collected at uniform time intervals $T_S = 1/f_S$, that is, at a sampling rate f_S, in which $f_S \geq 2f_M$. In fact, the condition for uniform time intervals is not necessary.

The sampling theory, derived by Claude E. Shannon, has been generalized for the case of non-uniform samples, that is, samples taken at non-equally spaced intervals (Davenport and Root, 1987).

It has been demonstrated that a band-limited signal can be perfectly reconstructed from its samples, given that the average sampling rate satisfies the Nyquist condition, independent of the sampling being uniform or non-uniform. (Margolis and Eldar, 2008).

For a band-limited signal $f(t) \leftrightarrow F(\omega)$, there is a frequency ω_M above which $F(\omega) = 0$, that is, that $F(\omega) = 0$ for $|\omega| > \omega_M$. Harry Nyquist concluded that all the information about $f(t)$, shown in Figure A.4, is contained in the samples of this signal, collected at regular time intervals T_S, as illustrated in Figure A.4. In this way, the signal can be completely recovered from its samples.

A.6 Sampling Theorem

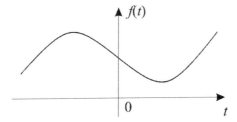

Figure A.4 A band-limited signal $f(t)$.

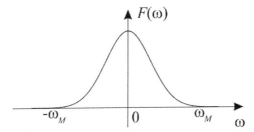

Figure A.5 Spectrum of a band-limited signal.

For a band-limited signal $f(t)$, that is, such that $F(\omega) = 0$ for $|\omega| > \omega_M$, as shown in Figure A.5, it follows that

$$f(t) * \frac{\sin(at)}{\pi t} = f(t), \text{ if } a > \omega_M,$$

because, in the frequency domain, this corresponds to the product of $F(\omega)$ by a gate function of width greater than $2\omega_M$.

The function $f(t)$ is sampled once every T_S seconds or, equivalently, sampled with a sampling frequency f_S, in which $f_S = 1/T_S \geq 2f_M$.

Consider the signal $f_s(t) = f(t)\delta_T(t)$, in which

$$\delta_T(t) = \sum_{n=-\infty}^{\infty} \delta(t - nT) \longleftrightarrow \omega_0 \delta_{\omega_0} = \omega_0 \sum_{n=-\infty}^{\infty} \delta(\omega - n\omega_0). \quad (A.80)$$

The periodic signal $\delta_T(t)$ is illustrated in Figure A.6. The Fourier transform of the impulse train is represented in Figure A.7.

The signal $f_S(t)$ represents $f(t)$ sampled at uniform time intervals T_S seconds. From the frequency convolution theorem, it follows that the Fourier transform of the product of two functions in the time domain is given by the

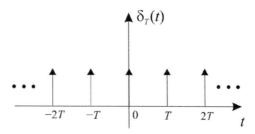

Figure A.6 Impulse train used for sampling the signal.

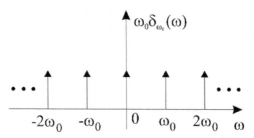

Figure A.7 The Fourier transform of an impulse train.

convolution of their respective Fourier transforms. It now follows that

$$f_S(t) \longleftrightarrow \frac{1}{2\pi}[F(\omega) * \omega_0 \delta_{\omega_0}(\omega)] \tag{A.81}$$

The important frequency convolution theorem implies that the Fourier transform of the product of two functions in the time domain is given by the convolution of their respective Fourier transforms. It now follows that

$$f_S(t) \longleftrightarrow \frac{1}{2\pi}[F(\omega) * \omega_0 \delta_{\omega_0}(\omega)] \tag{A.82}$$

and thus

$$f_S(t) \longleftrightarrow \frac{1}{T}[F(\omega) * \delta_{\omega_0}(\omega)] = \frac{1}{T}\sum_{n=-\infty}^{\infty} F(\omega - n\omega_0). \tag{A.83}$$

From Figures A.8 and A.9, it can be observed that if the sampling frequency ω_S is less than $2\omega_M$, the spectral components will overlap. This will cause a loss of information because the original signal can no longer be completely recovered from its samples. As the signal frequency ω_S becomes smaller than $2\omega_M$, the sampling rate diminishes causing a partial loss of information.

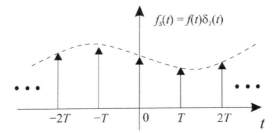

Figure A.8 Example of a sampled signal.

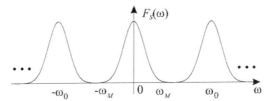

Figure A.9 Spectrum of a sampled signal.

Therefore, the minimum sampling frequency that allows perfect recovery of the signal is $\omega_S = 2\omega_M$, and is known as the Nyquist sampling rate, after Harry Nyquist (1889–1976), a Swedish engineer who made important contributions to communication theory. In order to recover the original spectrum $F(\omega)$, it is enough to pass the sampled signal through a low-pass filter with cut-off frequency ω_M.

If the sampling frequency ω_S is lower than $2\pi B$, in which B is the frequency in hertz, there will be spectra overlap and, as a consequence, information loss. As long as ω_S becomes lower than $2\pi B$, the sampling rate becomes lower, leading to partial loss of information. Therefore, the minimum sampling frequency for a baseband signal to be recovered without loss is $\omega_S = 2\pi B$, known as the Nyquist sampling frequency.

If the sampling frequency is lower than the Nyquist frequency, as mentioned, the signal will not be completely recovered, since there will be spectral superposition, leading to distortion in the highest frequencies. This phenomenon is known as aliasing, from the Latin word *alias*, meaning another, other, different.

If one increases the sampling frequency above the Nyquist rate leads to spectra separation that is above the minimum necessary to recover the signal, causing a waste of spectrum usage in the transmission.

A.7 Parseval's Theorem

For a real signal $f(t)$ of finite energy, the energy E associated with $f(t)$ is given by the following integral, in the time domain

$$E = \int_{-\infty}^{\infty} f^2(t)dt,$$

and can equivalently be calculated, in the frequency domain, by

$$E = \frac{1}{2\pi} \int_{-\infty}^{\infty} |F(\omega)|^2 d.\omega$$

Equating both integrals, it follows that

$$\int_{-\infty}^{\infty} f^2(t)dt = \frac{1}{2\pi} \int_{-\infty}^{\infty} |F(\omega)|^2 d\omega. \tag{A.84}$$

The relationship given in (A.84) is known as Parseval's theorem. For a real signal $x(t)$ with energy E, it can be shown, by using Parseval's identity, that the signals $x(t)$ and its delayed version, $y(t) = x(t - \tau)$, have the same energy E.

Another way of expressing Parseval's identity is as follows

$$\int_{-\infty}^{\infty} f(x)G(x)dx = \int_{-\infty}^{\infty} F(x)g(x)dx. \tag{A.85}$$

References

Alencar, M. S. and Rocha Jr., V. C. (2020). *Communications Systems*. New York: Springer.

Alley, C. L. and Atwood, K. W. (1973). Electronic Engineering, Third Edition. Wiley International Edition, John Wiley & Sons, New York.

Balanis, C. A. (2016). *Antenna Theory: Analysis and Design*. New York: Wiley-Blackwell.

Baskakov, S. I. (1986). *Signals and Circuits*. Mir Publishers, Moscow, USSR.

Campos, A. L. P. de S. (2015). *Laboratório de Princípios de Telecomunicações*. Rio de Janeiro: GEN.

Carlson, B. A. (1975). *Communication Systems*. McGraw-Hill, Tokyo, Japan.

Cutler, P. (1977). Circuitos Eletrônicos Lineares. McGraw-Hill, São Paulo.

Davenport, W. B. and Root, W. L. (1987). *An Introduction to the Theory of Random Signals and Noise*. Wiley-IEEE Press, New York, USA.

DeFrance, J. J. (1976). General Electronics Circuits, Second Edition. Holt RineHart Winston, New York.

Ding, B. P. (2010). *Modern Digital and Analog Communications Systems* (4 ed., Vol. 1). New York: Oxford University Press.

Farzaneh, F., Fotowat, A., Kamarei, M. Nikoofard, A. and Elmi, M. (2018). *Introduction to Wireless Communication Circuits*. Denmark: River Publishers.

Forouzan, B. (2012). *Data Communications and Networking*. New York: McGraw-Hill.

Fourier, J. B. J. (1888). *Théorie Analytique de la Chaleur*. Gauthier-Villars e tFils, Paris, France.

Gagliardi, R. M. (1988). *Introduction to Communications Engineering*. Wiley, New York.

Galup-Montoro, C. and Schneider, M. (2007). *MOSFET Modeling for Circuit Analysis and Design*, World Scientific Publishing Company.

Gomes, A. T. (1985). *Telecomunicações*. São Paulo: Érica.

Gronner, A. D. (1976). Análise de Circuitos Transistorizados. Livros Técnicos e Científicos Editora S.A., Rio de Janeiro.

Gummel, H. K. and Poon, H. C. (1970). An Integral Charge Control Model of Bipolar Transistors, The Bell System Technical Journal, vol. 49, pp. 827–852, May–June.

Haykin, S. (1988). *Digital Communications.* John Wiley and Sons, New York.

Haykin, S. (2001). *Communications Systems.* New York: John Wiley & Sons, Inc.

Kaufman, M. and J. A. Wilson (1984). Eletrônica Básica. McGraw-Hill, São Paulo.

Kolmogorov, A. N. and Formin, S. V. (1970). *Introductory Real Analysis.* Dove Publications, Inc., New York, USA.

Knopp, K. (1990). *Theory and Application of Infinite Series.* Dover Publications, Inc., New York.

Lathi, B. P. (2004). *Linear Systems and Signals* (Vol. 2). New York: Oxford University Press.

Lathi, B. P. and Ding, Zhi Modern (2010). *Digital and Analog Communications Systems.* New York: Oxford University Press.

Lee, T. H. (1998). *The Design of CMOS Radio-Frequency Integrated Circuits.* New York: Cambridge University Press.

Leon-Garcia, A. (2008). *Probability, Statistics, and Random Processes for Electrical Engineering.* New Jersey: Pearson Prentice Hall.

Lu, C. (1999). *Largura da Banda.* São Paulo: Berkeley.

Margolis, E. and Eldar, Y. C. (2008). "Nonuniform Sampling of Periodic Bandlimited Signals". *IEEE Transactions on Signal Processing,* 56(7):2728–2745.

Mayaram, D. O. (2008). *Analog Integrated Circuits for Communication* (Vol. 2). New York: Springer.

Millman, J. and Halkias, C. C. (1972). *Integrated Electronics: Analog and Digital Circuits and Systems,* McGraw-Hill Kogakusha, Ltd., Tokyo.

Nilsson, J. W. and Riedel, S. A. (2008). *Electrical Circuits.* New York: Prentice-Hall.

Oberhettinger, F. (1990). *Tables of Fourier Transforms and Fourier Transforms of Distributions.* Springer-Verlag, Berlin.

Padilha, A. J. G. (1993). Electrónica Analógica. McGraw-Hill, Lisboa.

Papoulis, A. (1983). *Signal Analysis.* McGraw-Hill, Tokyo.

Pederson, D. O. and Mayaram, K. (2008). *Analog Integrated Circuits for Communication.* New York: Springer.

Phillips, E. R. (1984). *An Introduction to Analysis and Integration Theory.* Dove Publications, Inc., New York, USA.

Riedel, J. W. (2008). *Electrical Circuits* (Vol. 8). New York: Prentice-Hall.

Sedra, A. S. and Smith, K. C. (2004). *Microeletronic Circuits*. New York: Oxford University Press.
Schwartz, M. and Shaw, L. (1975). *Signal Processing: Discrete Spectral Analysis, Detection, and Estimation*. McGraw-Hill, Tokyo.
Shichman, H. and Hodges, D. A. (1968). Modeling and simulation of insulated-gate field-effect transistor switching circuits, IEEE J. Solid-State Circuits. SC-3.
Silva, J. F. A. (2013). Electrónica Industrial, Semicondutores e Conversores de Potência. Segunda Edição Revisada e Actualizada. Fundação Calouste Gulbenkian. Lisboa, Portugal.
Smith, A. S. (2004). *Microeletronic Circuits* (Vol. 5). New York: Oxford University Press.
Smith, J. R. (1997). *Modern Communication Circuits*. New York: McGraw-Hill Science.
Stanley, W. D. (1982). Electronic Communications Systems. Reston Publishing Company, Inc. A Prentice-Hall Company. Reston, Virginia, USA.
Tsividis, Y. (1987). *Operation and Modeling of the MOS Transistor*, McGraw-Hill, New York.
Vervloet, W. A. (1978). Eletrônica Industrial. Livros Técnicos e Científicos Editora S.A., Rio de Janeiro.
Wozencraft, J. M. and Jacobs, I. M. (1965). *Principles of Communication Engineering*. JohnWiley & Sons, New York.
Wylie, C. R. (1966). *Advanced Engineering Mathematics*. McGraw-Hill Book Company, London.

Index

3 dB Bandwidth 87, 97

A

AC Compensation Capacitors 105
AC component 10, 13, 16, 23
Active Filters 85
Active Mixers 151, 159
Adder 47, 75, 99, 198
Adder Amplifier 47
AM Demodulator 201
amplifier 1, 42, 114, 215
non-inverter 46, 49, 60, 103
Non-inverting 43, 68, 118, 143
operational 1, 67, 162, 215
Amplifiers in Series 32, 33
Amplitude Modulation 184, 194, 206, 210
analog communications 163, 181, 184
angle 163, 207, 209, 210
Angular Modulation 160, 209, 210, 212
antiderivatives 220
Armstrong 123, 150, 213
Asynchronous 193, 194, 201
Augustin Louis Cauchy 220

B

Balanced Mixer 155, 156, 157, 158

Balanced Modulator 199, 200
Band-Pass 85, 87, 90, 96
Band-pass filter 87, 90, 96
Band-Stop 85, 87, 99
Band-stop filter 87, 99
bandwidth 24, 97, 209, 212
Barkhausen conditions 69
Barkhausen criterion 121, 128, 143
Baseband 181, 202, 212, 241
Behavior 16, 107, 140, 227
Bernhard Riemann 220
Bias Currents 56, 115, 119
Biasing 7, 17, 54, 114
Bilateral Exponential Signal 228
bipolar junction transistor 2, 16, 146
bipolar transistor 35, 38, 55, 114
BJT 2, 10, 17, 24
Bode diagrams 62
Bode plots 108, 110
bounded function 219
BPF 87, 97, 98, 99
BSF 87, 89
Buffer 71, 77, 112

C

capacitor 3, 70, 171, 204
capture range 165, 172, 174, 216
carrier 3, 163, 200, 217
carrier signal 184, 196, 198, 202

Carson Rule 212
central frequency 90, 99, 194, 214
channel 3, 181, 184, 194
characteristic equation 125, 130, 131, 143
characteristic function 130
closed-loop gain 31, 54, 69, 148
Coherent Detection 188
Colpitts oscillator 146, 149, 150, 202
common emitter 5, 14, 40, 195
Common Mode Rejection Ratio 48, 60, 103, 116
common source 5, 19, 23
common-emitter configuration 5, 19, 23
communications system 121, 163, 181, 208
Comparator 72, 151, 163, 180
complex coefficients 226, 227
Complex Exponential 231
components 1, 85, 160, 240
configuration 35, 119, 162, 216
Constant Function 230, 231
continuous spectrum 228
converter 78, 108, 152, 215
convolution theorem 237, 238, 239, 240
Cosine Function 231, 232
crystal 3, 69, 122, 123
current 1, 25, 119, 216
Current Amplifiers 28, 33
Cutoff Frequency 61, 90, 164, 216

D

Darlington configuration 35
DC voltage 162, 171

Demodulator FM Circuits 216
Demodulation 163, 188, 201, 218
demultiplexing 182, 183
deviation 208, 212, 214, 215
Differential Amplifier 35, 59, 60, 81
differential DC amplifier 43
Differentiation 21, 35, 235
Differentiator 76, 227
digital communications systems 163
digital radio 163
diode 42, 132, 151, 205
Dirichlet conditions 219
discontinuity points 219
Double Side Band 185, 190
downward conversion 152
DST 101, 105, 117, 119
Dynamic Operation 36, 38, 40

E

Early effect 6, 17, 21, 38
Ebers-Moll Complete Model 107
Edwin Armstrong 150
Edwin H. Colpitts 149
electronic circuits 13, 27, 122, 179
element 19, 70, 195, 205
Envelope 9, 185, 203, 218
Equivalent circuit 26, 38, 40, 123
Eudoxus of Cnido 220
Exponential Fourier Series 225, 226, 227, 232

F

Feedback amplifiers 29
feedback gain 30, 129, 148, 176
feedback loop 29, 31, 91, 142
Field-Effect Transistors 35, 55

Index 249

Filter Quality Factor 87, 92
Finite Gain Influence 48
First-order filters 88
first-order system 61
FM Modulator Circuits 213
FM Wave Direct Generation 215
FM Wave Indirect Generation 213
Forward current gain 16, 25, 26, 40
Fourier Analysis 220
Fourier Theory 219
Fourier Transform 187, 211, 228, 240
Linearity of the Fourier Transform 223
four-terminal 3
free-running mode 166
frequency 19, 105, 187, 242
Frequency Discriminator 216, 217, 218
Frequency Division Multiplexing 182
Frequency Domain Shift 235
Frequency Modulation 184, 207, 208
Frequency Response 29, 85, 118, 217
frequency synthesizer 163, 179, 180
Full-wave precision rectifier 80

G
Gain 12, 51, 112, 206
Gate Function 229, 239
gate transconductance 21
Gibbs phenomenon 220
Gottfried Wilhelm Leibniz 220
Gyrator 82, 83

H
half-wave precision rectifier 80
Harmonic Distortion 1, 128, 140
harmonic oscillator 122, 123, 126
Harry Nyquist 238, 241
Hartley oscillator 69, 146, 147, 149
Heinrich Georg Barkhausen 128
High-Impedance Differential Amplifier 35, 81, 82
High-Pass 85, 89, 95
High-pass filter 85, 89, 95
hybrid parameters 16, 25, 27, 29
hybrid-π model 17, 19
hysteresis 161, 162

I
Ideal Operational Amplifier 29, 42, 43
imaginary axis 130, 131, 143
Impulse Function 227, 230
inductive components 122
Influence of Gain and Impedances 51
Influence of temperature 59
Influence of the offset current 56
information 121, 181, 206, 241
Input bias current 114
Input impedance 12, 31, 85, 116
Input offset current 115
Input offset voltage 114
input signal 9, 161, 172, 238
Instability 29, 127, 129, 131
Instrumentation Amplifier 77, 81
Integration 35, 219, 236
Integrator 54, 76, 213
inversion 3, 35, 147, 191
Inverter Amplifier 45, 46, 67, 76
Inverter circuit 49, 50, 51, 60

Inverting Amplifier 67, 77, 89
inverting input 43, 44, 57, 118
Isaac Newton 220

J

Jean-Baptiste Joseph Fourier 220
jitter frequency 164
Joule effect 127

K

Kirchhoff's Voltage Law 124

L

lag 168, 179, 217
LC oscillators 121, 122, 123, 146
Lewis Method 101
limiter circuits 121, 131
Linear amplification 1, 25
linear transistor models 13
linearity 1, 13, 128, 233
LNA 33, 72
local oscillator 150, 188, 202, 215
locking range 165, 166, 174
Logarithmic Amplifiers 80
Loop of Tests 101
Loop Test 106, 114
lossless 124, 126
lossy resonator 126
Low Noise Amplifier 33
low pass filter 85, 169, 206, 241
lower limit 132, 136, 137, 174
Low-Pass 85, 164, 171, 241

M

macro-model 107, 110, 112
maximum gain 90, 94
maximum phase deviation 208
mesh gain 127, 131, 143, 178

message 152, 191, 210, 217
Method of Fluxions 220
Miller's theorem 52
mixer circuit 121, 155, 156, 202
modulation index 191, 212
modulation process 181, 191, 202, 218
Modulators Circuits 194
MOS transistor 2, 3
MOSFET 2, 12, 22, 106
multiple feedback circuit 94
Multiple Feedback Loop circuit 92
multiplexing 182, 183

N

Narrow-Band Angle Modulator 209
Negative Feedback 29, 45, 118, 164
negative resistance 122
Negative-feedback circuits 29
Noise Figure 33
Noise Filter 106
non-inverting input 43, 44, 57, 118
non-linear control loop 128
non-linear device 155, 196, 202, 206
Non-linear oscillators 121, 122
nonlinear resistor 109, 111
Nyquist Criterion 64
Nyquist sampling 241

O

Offset Adjustment 105, 118
Offset Current 48, 57, 102, 115
Offset Voltage 54, 101, 114, 119
OpAmps 42, 48, 67, 85

Operational Amplifier
  Real 48
Operational Amplifiers 27, 67, 110, 215
Orthogonality 220, 221, 222
Oscillators 69, 121, 146, 202
Output admittance 16, 25, 26, 40
output characteristics 5, 6
output impedance 12, 38, 85, 116
Output Swing 104, 105, 116
output wave 129, 135, 137, 144
overmodulation 191
overshoot 118

P

Parseval's Theorem 242
Passband 87, 118
Passive Mixers 151, 155
period 69, 121, 220, 239
Periodic Function 221, 227, 231, 232
periodic pulse train 198
periodic signal 121, 197, 227, 239
phase comparator 151, 168, 177, 180
Phase Modulation 184, 207
Phase-Locked Loop 160, 163
piezoelectric 123
plane of frequencies 130
PLL 163, 174, 179, 216
PLL Digital Circuit 179
Positive Feedback 44, 69, 122, 142
positive half-cycle 137, 150
power 1, 118, 193, 212
power amplifiers 27, 32
Power consumption 118
Power Supply Rejection Ratio 104, 116

Precision Rectifier 79, 80
Proportional and Quadratic Response 151, 154
pulsed 165, 169

Q

Quadratic Detector 202, 204
Quadratic Mixers 151, 153
Quadratic modulator 195, 196, 199, 204
Quadripole 59
quality factor 87, 92, 95, 98
quiescent point 11, 14, 18, 20

R

Radio 121, 150, 202, 230
radiofrequency systems 69
RC oscillators 122
RC Phase Shift Oscillator 69
receiver 179, 184, 189, 202
regenerative 122, 127
relaxation 69, 121, 123, 161
relaxation oscillators 121
resistors 2, 85, 131, 198
resonant tank 123, 146, 149, 150
Reverse voltage gain 16, 25, 26, 40
Riemann integral 220
ring structure 122
Rising time 118
roots 125, 143

S

Sallen-Key 91, 92, 95, 97
sampling frequency 239, 240, 241
sampling rate 210, 238, 240, 241
Sampling Theorem 210, 237, 238
sawtooth 121, 161
Scaling Property 223

252 *Index*

Schmitt Trigger 161
Second-order filters 90, 92
Series of Amplifiers 33
shift register 122
Shifter 77
Short-Circuit Current 105, 116
signal 1, 150, 181, 242
signal amplifiers 27, 32
signal analysis 219, 233
Signal-to-Noise Ratio 163, 194, 201, 216
sinusoidal oscillators 121, 126
Slew rate 107, 109, 117, 118
slew rate limitation 108, 109
small signal model 19, 22, 23
small signal operation 12, 19, 21, 23
small signals 2, 80, 104, 177
Small-signal equivalent circuit 38, 40
Source Resistors 106
spectrum 151, 181, 209, 241
square 83, 121, 162, 224
stability 35, 118, 127, 135
stability condition 128, 129
Static Condition 36, 37
Stopband 87
Subtractor 74, 75, 77, 81
superheterodyne receptor 201
Supply Current 105, 117, 118
Switching 106, 197, 198, 199
Symmetry Property 234
synchronism 163, 165, 189, 217
Synchronous 163, 194, 201, 205

T

The Compact Fourier Series 224
Thevenin's theorem 7, 57

Time Domain 207, 234, 237, 242
Time Domain Shift 235
Transconductance Amplifiers 28
Transimpedance Amplifiers 28
transistor base-emitter 197
transistors 2, 38, 116, 199
Transition band 87
transmitter 27, 181, 193, 194
Transresistance Amplifiers 78
triangular waves 121
Trigonometric Fourier Series 221, 223, 224, 227
tuned filter 202, 217, 218
tuned oscillators 121
two-port device 25, 26, 27
two-port network 16, 19, 110, 112

U

upper limit 24, 132, 139, 174
upward conversion 152

V

variation 3, 85, 161, 217
velocity 108, 182
voltage 1, 25, 136, 217
Voltage Amplifiers 27
voltage controlled current source 17, 109
Voltage-Controlled Oscillator 163, 164, 166

W

wavelength 182
Wide-Band Angle Modulator 210
Wien Bridge Oscillator 69, 70, 71, 122
Wien Oscillator 140, 143, 145, 146

About the Authors

Marcelo Sampaio de Alencar was born in Serrita, Brazil, in 1957. He received his Bachelor degree in electrical engineering, from the Federal University of Pernambuco (UFPE), Brazil, 1980, his Master's degree in electrical engineering, from the Federal University of Paraiba (UFPB), Brazil, 1988 and his Ph.D. from the University of Waterloo, Department of Electrical and Computer Engineering, Canada, 1993. He has 40 years of engineering experience, and 30 years as an IEEE Member, currently a senior member. Between 1982 and 1984, he worked for the State University of Santa Catarina (UDESC). From 1984 to 2003, he worked for the Department of Electrical Engineering, Federal University of Paraíba, where he was a full-time Professor and supervised more than 60 graduate and several undergraduate students. From 2003 to 2017, he was Chair Professor at the Department of Electrical Engineering, Federal University of Campina Grande, Brazil. He also spent some time working for MCI-Embratel and the University of Toronto, as a visiting Professor. He was visiting Chair Professor at the Department of Electrical Engineering, Federal University of Bahia.

He is the founder and president of the Institute for Advanced Studies in Communications (Iecom). He has been awarded several scholarships and grants, including three scholarships and several research grants from the Brazilian National Council for Scientific and Technological Research (CNPq), two grants from the IEEE Foundation, a scholarship from the University of Waterloo, a scholarship from the Federal University of Paraiba, an achievement award for contributions to the Brazilian Telecommunications

About the Authors

Society (SBrT), an academic award from the Medicine College of the Federal University of Campina Grande (UFCG), and an achievement award from the College of Engineering of the Federal University of Pernambuco, during its 110th-year celebration. He is a laureate of the 2014 Attilio Giarola Medal.

He published over 450 engineering and scientific papers and 24 books: Music science, Modulation Theory, Scientific Style in English, and Cellular Network Planning, by River Publishers, Spectrum Sensing Techniques and Applications, Information Theory, and Probability Theory, by Momentum Press, Information, Coding and Network Security (in Portuguese), by Elsevier, Digital Television Systems, by Cambridge, Communication Systems, by Springer, Principles of Communications (in Portuguese, by Editora Universitária da UFPB, Set Theory, Measure and Probability, Computer Networks Engineering, Electromagnetic Waves and Antenna Theory, Probability and Stochastic Processes, Digital Cellular Telephony, Digital Telephony, Digital Television and Communication Systems (in Portuguese), by Editora Érica Ltda, History of Communications in Brazil, History, Technology and Legislation of Communications, Connected Sex, Scientific Diffusion, Soul Hicups (in Portuguese), byEpgraf Gráfica e Editora. He also wrote several chapters for 11 books.His biography is included in the following publications: Who's Who in the World and Who's Who in Science and Engineering, by Marquis Who's Who, New Providence, USA.

Raphael Tavares de Alencar was born in Recife, Brazil, in 1988. He obtained his Bachelor Degree in Electrical Engineering Degree (Majors in Telecommunications and Power Systems), from the Federal University of Campina Grande, in 2012, and his Master's Degree in Electrical Engineering (Telecommunications) from the Federal University of Campina Grande, in 2014. He also got involved in an international exchange program at the École Nationale Supérieure des Systémes Avancés et Réseaux, (ESISAR), Valence, France, from 2010 to 2011.

He worked as a consulting engineer in a project for Alpargatas (manufacturer and distributor of Havaianas sandals in the country), and his main activities included development, manufacture, installation, and testing of a security system to protect mill operators. The main tasks involved electronic and mechanical designs, image processing, sensors, and radiofrequency systems testing. He also worked at the Laboratoire de Conception et d'Intégration des Systèmes (LCIS), in the development and manufacture of a chipless RFID reader.

He was a researcher in a Project funded by RNP (National Research Network): "[CIA]2 – Building Intelligent Cities: from Environmental Instrumentation to Application Development", developed at the Institute of Advanced Communications Studies (Iecom). His activities included mostly research, design, and simulation for optical communications and hybrid optical-wireless networks.

Raissa Bezerra Rocha was born in Campina Grande, PB, Brazil, in 1984. She received the B.E., M.S. and doctoral degrees in Electrical Engineering from the Federal University of Campina Grande, in 2010, 2012 and 2017, respectively. She is currently managing director of the Institute for Advanced Studies in Communications and an adjunct professor at the Federal University of Sergipe, where she teaches disciplines in the areas of communications and signal processing. Her main areas of interest are communications, voice processing, video, and biometric signals and communications circuits.

Ana Isabela Araújo Cunha was born in Salvador, BA, Brazil, in 1966. She received the B.E. degree in Electrical Engineering from the Universidade Federal da Bahia (UFBA), Salvador, in 1989, and the M.S. and doctoral degrees in Electrical Engineering from the Universidade Federal de Santa Catarina, Florianópolis, SC, Brazil, in 1993 and 1996, respectively. In 1995, as part of her doctorate, she spent one year at École Nationale Supérieure d'Electronique et Radioélectricité de Grenoble, France. Since 1990, she has been with the Electrical Engineering Department at UFBA, where she is a professor. In 2005 she spent a one-year sabbatical at the Department of Electrical Engineering and Computer Science at the University of California Berkeley, USA. Her research interests include metal–oxide–semiconductor field-effect transistor modeling and analog integrated circuit design.